XXIII CONGRES INTERNATIONAL DE GEOGRAPHIE
XXIII INTERNATIONAL GEOGRAPHCAL CONGRESS
XXIII МЕЖДУНАРОДНЫЙ ГЕОГРАФИЧЕСКИЙ КОНГРЕСС
URSS*USSR*CCCP* MOSCOU*MOSCOW*МОСКВА*1976

INTERNATIONAL GEOGRAPHY
GEOGRAPHIE INTERNATIONALE
МЕЖДУНАРОДНАЯ ГЕОГРАФИЯ

'76

Additional Volume

Volume supplementaire

Дополнительный том

12

MOSKVA 1976

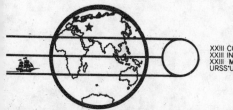

XXIII CONGRES INTERNATIONAL DE GEOGRAPHIE
XXIII INTERNATIONAL GEOGRAPHCAL CONGRESS
XXIII МЕЖДУНАРОДНЫЙ ГЕОГРАФИЧЕСКИЙ КОНГРЕСС
URSS*USSR*CCCP*MOSCOU*MOSCOW*МОСКВА*1976

INTERNATIONAL GEOGRAPHY
GEOGRAPHIE INTERNATIONALE '76
МЕЖДУНАРОДНАЯ ГЕОГРАФИЯ

Additional Volume

Volume supplementaire

Дополнительный том

Distributed by PERGAMON PRESS LTD., Headington Hill Hall, Oxford
OX3 0BW, England
Also New York, Sydney, Toronto, Paris, Frankfurt
British Library Cataloguing in Publication Data
International Geographical Congress, *23rd Moscow 1976*
International geography '76
1. Geography — Congresses
I. Title   II. Gerasimov, Innokentii Petrovich
ISBN   Set 0 08 023154 3   Volume 12   0 08 023152 7
Printed in the U.S.S.R.

MOSKVA 1976

EDITOR-IN-CHIEF   ACAD. I.P. GERASIMOV
EDITEUR-EN-CHEF

ОТВЕТСТВЕННЫЙ
РЕДАКТОР

EDITORS OF THE VOLUME   ASEEV A.A.
EDITEURS DU VOLUME   VORONOV A.G.
РЕДАКТОРЫ ТОМА   GOKHMAN V.M.
                         KOVALEV S.A.
                         PREOBRAZHENSKY V.S.
                         PRIVALOVSKAYA G.A.
                         KHOREV V.S.
                         TSYGELNAYA I.D.

SECRETARY   SEREGINA M.Yu.
SECRETARY
СЕКРЕТАРЬ

                    Переводы на иностранные
                    языки представлены
                    Оргкомитетом конгресса.

Preface

This volume represents one of the issues of the series
"International geography, 1976" in which materials for sections, general symposia and methodical seminars of the XXIII
International Geographical Congress are published. Section
papers are published in the following volumes:

1. Geomorphology and Paleogeography
2. Climatology, Hydrology, Glaciology
3. Geography of the Ocean
4. Biogeography and soil geography
5. General Physical Geography
6. General Economic Geography
7. Geography of Population
8. Regional Geography
9. Historical Geography
10. Education of Geography, Geographical Literature
    and Dissemination of Geographical Knowledge.

Short papers for general symposia and methodical seminars are published in volume 11 - "General Problems of Geography and Geosystems Modelling".

The present volume contains the additional materials of
the Moscow programme which have not been included in sections
1 and 2 for technical reasons.

Reports are arranged according to the thematical principle - the number of sections.

The volume has a name index for all 12 volumes. Referen-
ces are made to the volume (Roman figures) and to the page
(Arabic figures).

Preparing the volumes of "International Geography, 1976" the Programme Commission of the Organizing Committee was guided by the requirements to the papers defined in the Second Circular: accordance with the Programme of the Congress, originality and scientific value of broad interest, conformity of presentation with the requirements for publication and observation of the deadline for receipt of the text.

The limit of about 1000 words for a publication has allowed to the authors to present the main idea of the paper what can be a basis for a scientific discussion at sessions.

The editors of the volumes limited their work by the unification of the presentation of papers and by the shortening of those papers which largely passed over 1000 words.

The papers are published in one of the official Congress languages as submitted by the authors.

Papers are placed according to the subject principle and in the alphabetic order.

Préface

Ce volume est un des plusieurs volumes qui forment la série "Géographie Internationale, 1976" dans laquelle on a réuni les matériaux préparés pour les sections, les symposiums généraux et les séminaires méthodiques du 23e Congrès Géographique International. Les rapports présentés pour les sections sont publiés dans les volumes suivants:

1. Géomorphologie, Paléogéographie
2. Climatologie, Hydrologie, Glaciologie
3. Géographie des océans
4. Biogéographie et géographie du sol
5. Géographie physique générale
6. Géographie économique générale
7. Géographie de la population
8. Géographie régionale
9. Géographie historique
10. Enseignement géographique, publications géographiques, diffusion des connaissances géographiques

Le volume XI - "Problèmes généraux de la géographie et de l'établissement des modèles mathématiques des géosystèmes" - contient les textes abrégés des rapports présentés aux symporiums généraux et séminaires méthodiques.

Nous publions dans le présent volume les matériaux supplémentaires du programme de Moscou qui n'ont pas paru dans les fascicules 1-11 pour des causes techniques.

Les rapports sont classés selon le principe thématique, dans l'ordre des numéros des sections (des fascicules).

Ce volume comprend l'index des auteurs par ordre alpha-
bétique pour toute la série des 12 volumes. Le numéro du
volume est en chiffre romain et celui de la page en
chiffre arabe.

En préparant les volumes de la "Géographie Internationale, 1976" la Commission de programme basait son travail
du Comité d'organisation sur les exigences concernant les
rapports indiquées dans la deuxième Circulaire: conformité
avec le programme du Congrès, originalité, actualité et un
large intérêt du contenu scientifique, présentation du texte
en accord avec les instructions pour la publication et observation des délais établis.

La limite de près de 1000 mots établie pour un rapport
a permis aux auteurs de présenter le contenu de leurs communications sous une forme qui peut servir de base pour la discussion scientifique lors des séances.

Le travail des rédacteurs de chaque volume s'est limité
à l'unification de la présentation des textes et à l'abrège-
ment de certains textes qui largement dépassaient les
1000 mots.

Les rapports sont publiés dans une des langues officiel-
les du Congrès tels comme ils ont été présentés par les
auteurs.

Les rapports sont systématisés dans l'ordre alphabéti-
que et groupés par les thèmes.

ПРЕДИСЛОВИЕ

Данный том представляет один из выпусков многотомной серии "Международная география - 1976", в которой публикуются материалы, подготовленные для секций, общих симпозиумов и методических семинаров XXIII Международного географического конгресса. Секционные доклады публикуются в следующих томах:

1. Геоморфология и палеография
2. Климатология, гидрология, гляциология
3. География океана
4. Биогеография и география почв
5. Общая физическая география
6. Общая экономическая география
7. География населения
8. Региональная география
9. Историческая география
10. Географическое образование, географическая литература и распространение географических знаний.

Краткие тексты докладов на общих симпозиумах и методологических семинарах составляют II том "Общие проблемы географии и моделирование геосистем".

В данном томе публикуются дополнительные материалы московской программы, не вошедшие в выпуски I - II по техническим причинам.

Доклады расположены по тематическому принципу - по порядку номеров секций(выпусков).

Этот том включает в себя алфавитный авторский указатель для всей (12-ти томной) серии: индекс указывает номер тома (римские цифры) и страницу(арабские цифры).

При подготовке выпусков "Международная география - 1976". Программная комиссия руководствовалась, насколько это было возможно, при столь широком международном Конгрессе, требованиями к докладам, определенными вторым циркуляром: соответствие программе Конгресса; оригинальность, актуальность, научная содержательность и широкий интерес; оформление текста в соответствии с формой, необходимой для публикации и соблюдение сроков присылки. Принятый для публикации размер около 1000 слов позволил авторам изложить основное содержание доклада, которое может служить основой для научной дискуссии во время заседаний.

— При подготовке материалов к опубликованию ответственные редакторы соответствующих томов проводили лишь минимально необходимую правку текстов, не влияющую на стиль автора, а в случаях значительного превышения объема (1000 слов) - сокращение текста.

Доклады печатаются на одном из официальных языков Конгресса по оригиналу, представленному автором.

Доклады расположены в алфавитном порядке по тематическому принципу.

I. GEOMORPHOLOGY AND PALEOGEOGRAPHY

Bandyopadhyay, M.K.
(INDIA, Calcutta)

DRAINAGE DEVELOPMENT IN THE LADAKH HIMALAYA

The district of Ladakh occupies the eastern part of Kashmir. The area is drained by the Indus and its tributaries. On the northern slope of the Kailas Range in southern Tibet originates the Sengge Khabab or the Indus which, in its upper course, flows at first to the north and then gradually turns to the west to join the Gartang river near Tashigong and then, taking the direction of the Gartang, enters the state of Kashmir through its southeastern corner, at about 20 km from Tashigong.

Crossing the Ladakh Range: — On entering Kashmir the Indus follows closely the northeastern slope of the Ladakh Range. Here the Ladakh Range has a northwest-southeast trend and the river also flows parallel to it keeping the range on its left. After flowing for about 120 km from the confluence with the Gartang, the Indus takes a sharp left-hand turn near Lung Kung (33°12' N, 78°58' E) and crosses the Ladakh Range from the northeast to the southwest. At the right-angle bend near Lung Kung the Indus flows to the southwest for about 5 km and then again turns to the northwest and flows parallel to the Ladakh Range keeping the Range on its right.

The Zaskar River: – Beyond Lung Kung there are several tributaries from both sides of the Indus of which the Zaskar is of considerable volume and length. It originates in the small cirque glaciers of the Lahul Himalaya and flows in a zig-zag way with numerous right-angle bends which suggest structure control. The general trend of the river in its upper course is in the di-

9

reotion of the strike, and in the lower portion, across the strike, of the mountain ranges. During field work in the area it was observed that the volume of discharge of the Zaskar was considerably greater than that of the Indus. This was because of the snow-melt water from the numerous glaciers of the Zaskar Range which fed the Zaskar River in summer. Also the water in the Zaskar was considerably muddy as compared to that of the Indus.

The Suru River Basin: - In the westernmost part of the Indus basin in the Ladakh district lies the Suru-Shingo River basin. The Suru river originates in the glaciers of the Nunkun mountain in the Zaskar Range and drains the northern and the eastern slopes of it. The upper part of the Suru near the eastern foot of the Nunkun is known as the Sankpo which flows to the north; next it turns to the west and passes by the northern foot of the mountain. The middle portion of the stream is known as the Karcha while the lower portion is called the Suru.

The Sankpo i.e., the upper part of the Suru, flows to the northeast i.e., across the strike of the Himalayan axis in the area. It appears that this portion of the river originated on the initial slope of the mountain fold and flowed down the initial slope. On the contrary, the portion next to it is flowing in the direction of the mountain fold i.e., from the east to the west along the strike, and hence, may be regarded as a strike stream. The lower portion of the Suru river, however, crosses several folds in succession and is much more complex in character. Here also the minor tributaries join the main river following the direction of the strike. The zig-zag course of the Suru river and its course across the Zaskar group of ranges needs more serious attention. Probably the river is antecedent to the Zaskar Range, i.e., it is older than the mountain range it crosses, and has been able to maintain its course across the mountain range as the mountain rose gradually across its way along with the general rise of the Himalaya.

The Shingo Basin: Signs of Recent Glaciation: - The Shingo originates near the western border of the Ladakh district and flows across the general direction of the strike of the mountain ranges in the area. The most important tributary in the left is the Shigar River which, however, flows parallel to the strike of the mountain ranges.

The Dras River on the other hand, originates in the south, crosses the mountain ranges in between, and joins the Shingo near Kargil as a right-bank tributary.

There are some interesting differences in the character of the two systems:

First, the whole area shows marks of past glaciation: scattered moraines, erratics, gently rolling, topography, U-shaped valleys, etc. are evidences of that. However, the southern part shows more recent glaciation of the Little Ice Age.

Secondly, the Shingo, the Shigar and several other tributaries which originate in the Ladakh Range and its spurs in the north, do not have any glacier at their source today. On the other hand, the Dras and its tributaries have several glaciers, though small in size, on the northern slope of the Zaskar Range.

A possible explanation of the above phenomenon is that in the south the north-facing slope of the Zaskar Range protects the glaciers from direct sunshine whereas the south-facing slope of the Ladakh Range in the north does not bear any glacier today because of the scorching rays of the sun. Also it appears that the snow fall during the Little Ice Age was not enough to create new glaciers on the Ladakh Range in the north although it was sufficient to feed the already existing glaciers of the Zaskar Range in the south.

The combined volume of the Shingo, the Shingar and the Dras joins the Suru river a few kilometres to the north of Kargil and then flows to the north-east to join the Indus as a left-bank tributary.

The Right-Bank Tributaries of the Indus: General Characteristics: - The Shyok is the only important right-bank tributary of the Indus in the eastern part of the State of Kashmir. Most of the tributaries which drain the southwestern slope of the

Ladakh Range are not at all significant so far as their length
or the volume of discharge are concerned. This is mainly because of the fact that the Ladakh Range does not contain any important glacier especially on its southwestern slope. Also the
rainfall in the area is very low. Moreover, the summit of the
Ladakh Range which forms the divide between the Indus and the
Shyok, is not far from the channel of the Indus and the slope
is quite steep. Above all, the area is almost devoid of any
kind of vegetation and the barren surface does not give any
resistance to the surface flow of water. As a result, most of
the streams have very low discharge throughout the year whereas sudden cloud outburst may sometimes cause severe flood damage in the lower reaches of such rivers.

The Shyok: - The Shyok, on the other hand, is fed by a large
number of important glaciers of the Karakoram Range which enables it to maintain a considerable volume of discharge throughout the year. In its upper course one of the main sources comes from the Rimo glacier which is about 40 km long. The river
here flows in a south-easterly direction. Just above Shyok, a
village on the Shyok river, quite close to the Pangong Tso,
the river turns abruptly to the northwest. This abrupt change
of course by about 170° indicates that the river in this area
is dominantly controlled by the structure. The Nubra, a tributary to the Shyok, and the Shigar II, a tributary to the Indus,
also show similar obtuse-angle bends in the lower portion.
Both above and below the bend the river flows along the strike
of the ranges; only a small portion of the river near the bend
lies across the Karakoram Range. The river was able to follow
such a course possibly because of some weakness across the direction of the strike in the area. The striking alignment of
the Pangong Tso (a narrow but elongated lake) with the upper
parts of the Shyok and the Indus may, at a first glance, suggest their intimate relationship, but a minute examination of
the relief in between the lake and the rivers concerned will
reveal that they never formed parts of the same system. Such
alignments of the drainage features appears to be rather a
reflection of the alignment of the tectonic depressions formed by the initial folding.

Conclusion: -

The above study indicates that:(a) the Indus is antecedent to the Ladakh Range, (b) the structure has a dominant
control on the course of the Indus proper, as well as on its
major tributaries, (o) significant influence of the aspect on
the intensity and extension of glaciation is evident, and that
(d) the obtuse-angle confluence of several streams with the
master stream does not suggest any reversal of the direction
of flow of the Indus but simply suggests structure control.

Buzek L.
(ČSSR, Ostrava)

THE RELATION OF EROSION BY RUNNING WATER TO HYDROMETEOROLOGICAL
SITUATIONS AND DISSECTION OF RELIEF (ON THE EXAMPLE OF SMALL
DRAINAGE BASINS IN THE NORTHERN FORELAND OF THE FLYSCH CARPA-
THIANS

Soil as a very complicated dynamic system is destroyed by
exogenic factors and in many cases it is entirely degraded by
the economic human activity. The main factor of degradation of
soil fund in the humid areas is running water. In dissected relief affected by human activity (with the help of rain or melting water) it takes away a great amount of solid and dissolved
materials from its drainage basin (N.M. Strakhov, 1962). This
material was carried into river channels by sheet erosion from
steeper slopes without vegetation or from slopes with agricultural vegetation with a very low antierosional influence (esp.
potatoes or beet) in the time of heavy rains. The preventive
function of vegetation is widely known (S.S. Sobolev, 1960,
T. Gerlach, 1966, D. Zachar, 1970, H. Hundson, 1971), but human activity does not respect this protective function in all
cases.

The intensity of erosion by running water in the Flysch
Carpathians varies according to relief, the manner of agricultural land-use and the character of rains and their intensities.
In the years 1971 - 1974 the erosion by running water was studied in the area of the Flysch Carpathians Foreland in North-

13

ern Moravia, esp. in the Podbeskydské Hillylands and the western part of the Radhoštská Mts. In the first stage of investigation there were reckoned amounts of carried materials on the basis of natural factors (potential soil erosion) and on the basis of combined natural and anthropogenic factors (probable soil erosion see O. Stehlík, 1970, 1971). The investigated area has in the northern part a character of hillyland with very low relative heights (up to 100 m) and low declivities of slopes (predominantly to 10°), but the southern part of the area has a highland or mountainous character with relative heights to 500 m and declivities of slopes over 16°. The flysch substratum with a great amount of claystone (Cretaceous and Paleogene systems) and the sediments of the loess loams in the northern part with intensive agricultural activity in the valleys and basins and thinly forested areas (to 30%) enable the carrying away of fine soil particles.

The direct measurement of suspended matter which expresses the erosional processes in the studied area was carried out in three drainage basins with dissimilar relief and size.

Drainage basin	Surface in sq. km	The character of relief	The area of forests in percentage of the area of drainage basin
Ondřejnice, upper part	41.6	highland	30
Ondřejnice, lower part	57.8	dissected hillyland	20
Ondřejnice, the whole basin	99.4	highland and dissected hillyland	25
Lubina	163.4	highland and mountainous relief	30
Jasénka	16.2	indissected hillyland	10

The water for the analysis of suspended matter was taken in certain profiles every day and in the case of extraordinary meteorological situation more times during one day. The extraordinarily dry period of 1973 - 1974 showed us a very direct relations of suspended load to rainy situations, because during short intensive rains the amount of suspended matter grew up enormously, as compared with days without precipitations. In the investigated period, there were in the studied area the precipitations lower by 25 - 30% as compared with the long-term average precipitations (50 years) but occasional torrential rains in summer and snowmelt waters demonstrate intensive relief processes according to the amount of suspended matter; with the exception of precipitations we must regard the character of relief and the density of vegetation too.

The enormous growing of suspended load in the rivers after the precipitations at the end of July 1973 is demonstrated by the following table:

Basin	Date	Precipitations in mm	Discharge of water in m^3/s	Discharge of suspended matter in g/s
Ondrejnice,	27.7	19.7	0.9	37.8
upper part	28.7	33.7	5.3	276.6
Ondrejnice,	27.7	20.4	1.9	792.5
the whole	28.7	31.5	7.1	2380.6
basin				
Lubina	27.7	23.5	0.9	43.7
	28.7	31.5	5.3	5229.0
Jasénka	27.7	3.8	0.035	0.5
	28.7	34.8	0.08	17.1

Analogously in the time of the melting of snow which was in many cases connected with rain precipitations we observe an intensive carrying away of solid material because the soil is not protected by vegetation. It is possible to illustrate this fact on the example of the drainage basin of the Lubina R., when in January 1974, in the course of 5 days nearly 92% of the whole amount of solid matter of this month was carried away (the melting of snow and heavy rains); at the same time in the low dissected relief of the Jasénka Brooklet basin 85% of the whole January amount was carried away too.

The most rainly months in the years 1973 - 1974 were July (storms), December and January (the melting of snow and rains) and during this quarter of the year in the soil erosion was most intensive.

Months	Suspended load in the percentage of the year amount				
	Lubina	Upper part Ondřejnice	Lower part Ondřejnice	The whole basin Ondřejnice	Jasénka
July+ December+ January	69.45	64.93	79.12	71.17	56.52

Not only precipitations and their intensities but also the dissection of relief in the investigated area influenced the amount of suspended matter. If precipitations were greater than 1 mm, the carrying away of solid material in tons per sq. km during one year in single basins was following:

The basin of the Lubina R. .... 16.4 t/sq.km
The upper part of the basin of the
   Ondřejnice R. .... 15.0 t/sq.km
The whole basin of the Ondřejnice R. .... 13.3 t/sq.km
The basin of the Jasénka Brooklet .... 2.4 t/sq.km

In all investigated basins during the precipitations of 1 mm and more and during the melting of snow 81.6% solid material of the year amount was carried out, which represents the wasting of 0.002 - 0.01 mm of soil during one year. This relatively low amount corresponds only to 80 days with precipitations, so that a year with an average amount of days with precipitations or with an abnormal amount of such days must have greater erosion through running water. For a similar relief

A. Reniger (1957) mentions losses of 0.052 - 0.154 mm during one year and A. Tłałka (1967) 0.027 mm/year.

The contemporaneous erosion by running water in the investigated area does not seem to be great and it is possible to suppose that the contemporary antierosional arrangement (esp. old agricultural terraces) protect the soil sufficiently. To prevent a greater destroying of the soil it must be urgent to observe the agrotechnical principles of cultivating the friends and the suitable pattern of agricultural vegetation. It is not possible to admit in any case the lowering of forested areas because the trees have an enormous protective function for soil, esp. on the flysch substratum.

Czudek T.
(ČSSR, Brno)

NEW GEOMORPHOLOGICAL DIVISION OF THE RELIEF OF THE CZECH SOCIALIST REPUBLIC

The geomorphological regionalization of the Czech Socialist Republic was finished in the Institute of Geography of the Czechoslovak Academy of Sciences in recent years by a working team headed by the author of this report. The regionalization consists of two parts mutually closely connected, the regional and the typological divisions of the relief. The geomorphological units of the western part of Czechoslovakia were represented on colour maps at a scale 1:500 000. In this report the regional division will be discussed.

The task of the regional division of the relief of the Czech Socialist Republic was to define geomorphological units of various taxonomic degrees and incorporate them in the regional system. The delimitation of these units was based on morphology, morphostructure, genesis and development of the relief. The morphographical characteristic was based on relief amplitude (relative relief) derived from maps at a scale 1:25 000 on an area of 16 km^2 with overlapping 1 cm^2 areas. The result was the so-called dasymetric map allowing to cha-

I7

racterize the geomorphological units as follows: in the case
of a relief amplitude between 0 and 30 m as plains, 30 and 75 m
as less dissected hilly lands, 75 and 150 m as more dissected
hilly lands, 150 and 200 m as less dissected highlands, 200
and 300 m as more dissected highlands, 300 and 450 m as less
dissected mountains and 450 and 600 m as more dissected mountains. The bottoms of basins and hollows usually have the character of plains or hilly lands. In dividing of geomorphological units into the morphographical relief classes mentioned
above the prevailing relief amplitude was decisive i.e. that
occurring in the territory in question on as much as 80 per
cent of its area or typical of the territory. In the interpretation of the results some disadvantages of the cartographical
method applied were compensated by taking into consideration
the altitudes and the mutual relations between neighbouring
geomorphological units.

The basic unit in the regional division of the relief of
the Czech Socialist Republic is the geomorphological region
comprising usually several relief types (the same relief type
occurs in several regionally defined geomorphological units).
It is a geomorphological unit in a certain regional position
associating geomorphological subregions of the same morphostructure and/or relief genesis and development, distinctly
differing from the adjacent geomorphological regions. The borders between regions are formed mainly of slope foots. Geomorphological regions join into subsystems defined predominantly
on the basis of a uniform morphostructure and similar morphographical conditions. The subsystems join into geomorphological systems corresponding prevailingly to basic morphostructural units of the western part of Czechoslovakia. The systems
join into provinces corresponding to structural tectonic units
of higher order (macromorphostructures) such as the Bohemian
Massif, the Western Carpathians. The geomorphological regions
are divided into subregions characterized by similar features
as regions but displaying a greater morphographical and genetic relief homogeneity.

According to prevailing relief amplitude, morphostructure, relief genesis and development as well as with respect
to the geomorphological divisions of adjacent countries 368
geomorphological units were delimited on the territory of the
Czech Socialist Republic. These are 4 geomorphological provinces, 10 systems, 27 subsystems, 93 geomorphological regions
and 234 subregions. For these units their highest and lowest
points, area, mean altitude and average slope were established.

The geomorphological division of the relief of the Czech
Socialist Republic was discussed with many Czechoslovak and
foreign geomorphologists. After a public discussion it has
been accepted and the names of the geomorphological units were
approved by the Terminological Commission of the Czech Board
for Geodesy and Cartography in Praha. Thus they came into force
and are now currently used in schools, popular and scientific
publications, etc.

Del Busto, R.; Iñiguez, L.
(CUBA, Havana)
PECULIAR TRAITS OF CUBAN RELIEF

Among the unique manifestations which characterize the
development of Cuban relief the following can be distinguished
by the originality and extension of its distribution: the surfaces with different types of karst, the presence of anticlinal depressions, the Plio-Quaternary marine terraces and thick
weathering crusts.

Cuba is a country with a great variety of karst forms and
types. Over 50% of the territory is affected by karst processes. Particularly in Havana and Matanzas provinces, the morphostructural pattern which has directed the development of the
morphogenesis during the post-Miocene epoch was created by the
combination of an Oligo-Miocene crust, generally carbonated,
dislocated in folded block massif ranges and in slightly folded

monoclinal plains, with a system of latitudinally originated anticlinal valleys where sedimentary and ultrabasic (mainly Mesozoic) rocks of insoluble nature are on the surface.

The karst, karst-denudation and denudation processes carved the highest surfaces and in a lesser degree the Quaternary marine terraces. The karstified territories of Havana province occupy 60 - 70% of the total area, coinciding particularly with the zones of greatest agricultural production, population density and most important localities of water supply. The karst typology reflects morphogenetical and geological conditions which have played the principal role in the differentiation taking into account basically the degree of karstification, the thickness and type of covering material and the morphostructural and lithological conditions of its development. Two principal categories were thus determined: karstified heights and plains. These were subdivided into 16 types, each of which possesses different availability for exploitation (agriculture, construction, water extraction) and therefore they require specific treatment in the general activities of the organization of space. (See map).

In the middle northern part of Havana and Matanzas provinces, a typical inverted relief phenomenon is present: the anticlinal valleys which occupy areas from 30 - 40 km long and 10 - 12 km wide approximately. The progressive wearing away of the top of this structure shows the intensity of the exogenetic post-Miocene processes in favourable lithological structural conditions.

In the interior of these inverted morphostructures there appears a real lithological kaleidoscope, composed of flysch and marl type rocks which give rise to a hilly relief (60 - 80 m) of rounded tops and convex slopes with characteristic phenomena of solifluction which responds principally to anthropogenic activity and high humidity indexes. This hilly relief is interrupted by long low ranges (100 - 200 m) composed of serpentine rocks with NW-SE direction coinciding with the ancient axes of Cretaceous structures and discordant with the

20

latitudinal direction of the recent folds. These ranges constitute unearthed geological bodies which are resistant to the
action of denudation processes.

The fluvial systems within the anticlinal depression
have, in occasions, produced relatively extensive fluvial valleys, with systems of alluvial terraces which are well developed, where the exorheic drainage was attained through the
formation of deep gaps which dissected in different degree
the monoclinal carbonate rocks which surround these valleys
(Yumurí, Guanabo rivers).

The Plio-Quaternary eustatic changes made possible on the
Cuban coasts the development of an extense and vigorous system
of marine terraces which left a characteristic mark creating
a layered structure with different degrees of conservation.

This layered structure is developed basically to the
height of 90 - 100 m with eight principal levels of marine
plantation.

In some places (Maisí and Cabo Cruz) a great number of
terrace levels (15 - 18) are found to heights of over 400 m
as a result of the continuous neotectonic movements.

The terraces are well preserved only in those places
where they have been carved mainly in Miocene limestone, while
in localities of other lithological types these have been
transformed by erosive and slope processes. Generally the low
terraces (5-6, 10-12, 20-25) present well defined scarps and
niches.

The nickeliferous laterite weathering crusts of the
North-Eastern mountain massif of Oriente province are developed on ultrabasic massifs. There areas contain Cuba's most
important mineral resources.

They represent high plains and remains of ranges which
reach heights of 800 - 900 m dissected in their marginal parts
and composed of a system of polygenetic planation surfaces.
The height and dissection of the high plains created optimum
conditions for the circulation of subterraneous waters which
facilitated the formation of thick weathering crusts, under
conditions of a seasonally humid tropical climate. Although

the crusts have been formed on planation surfaces of different ages and heights, their formation can be considered as a
continuous process.

Morphometric studies in typical localities showed relations between different crust parameters (nickel and iron content) and the variable elements of the morphology (slope,
height and dissection). These studies also showed the inter-
relation between the crusts parameters themselves, with the
following conclusions:

a) The crusts are well developed on relatively steep slopes, since the greatest thicknesses (5 - 12 m) are found on
slopes 15 - 20°.

b) Where the thickness of the crust is greatest (5 - 12 m)
the iron content is maximum (45%), while the nickel content is
notably greater (1.4 - 2.0%) for thicknesses of between 0 and
5 m.

The traits of Cuban relief which are analyzed are the result of the complex interaction of zonal and azonal factors
which complicate extraordinarily the geomorphological structure of the country.

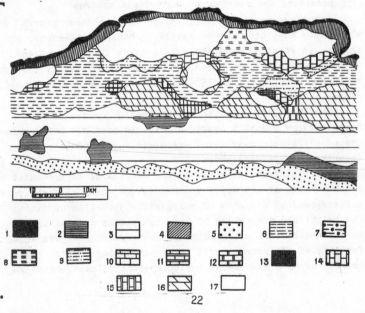

22

MAP OF HAVANA PROVINCE KARST

1. KARSTIFIED PLAINS. A. On monoclinal structures
   mainly.
1 - Coastal Plain with Naked Karst on Quaternary Limestones; 2 - Abrasive Plain with Naked and Semi-
covered Karst on Miocene Limestones; 3 - Abrasive
Plain with Karst predominantly covered by Red Soils on
Miocene Limestones; 4 - Terrace Plain with Semi-
covered Karst on Miocene Limestones; 5 - Marine -
Swampy Plain with Karst covered by Marl-Clay sediments
and Turf fragments on Miocene Limestones. B. On folded
structures. 6 - Abrasive-Denudative Plain with Karst
covered by Red Soils and Weathering Crust on Miocene
Limestones with synclinal   structure mainly; 7 -
Abrasive-Denudative Plain with Karst covered by Red
Soils with Cupula Karst on Miocene Limestones of
synclinal structure; 8 - Abrasive-Denudative Plain
with Karst covered by Red Soils on top of a gentle
anticline made of   Miocene Limestone. 9 - Fluvial
Plain with Karst covered by   alluvial sediments and
Weathering Crust on Miocene Limestones of synclinal
structure. II. KARST HEIGHTS. A. On block folded
structures. a. Plateaus with Karst covered by Red
Soils   on Oligo-Miocene Limestones. 10. On horst-
synclinal structure; 11 - On block periclinal structure; 12 - On block-anticlinal structure. b. Cone
Karst on Oligo-Miocene Limestones. 13 - On monoclinal structure. c. Dome Karst on Oligo-Miocene Limestones. 14 - On synclinal structure; 15 - On monoclinal structure. d. Heights made of sedimentary rocks
(Marl Limestone, Marl, and others) with very slightly
developed Karst. 16 - On block-anticlinal and monoclinal structure. III. OTHER SYMBOLS. 17 - Non-Kars-
tified Territories.

23

Hrádek M., Ivan A.
(ČSSR, Brno)

THE MORPHOSTRUCTURES OF THE EASTERN MARGIN OF THE ČESKÁ VYSOČINA (BOHEMIAN HIGHLANDS)

The eastern margin of the Česká Vysočina (Bohemian Highlands) was the contact region of the Epivariscan platform and the Alpine-Carpathian geosynclinal region in Mesozoic and Tertiary and this position influenced strongly its geological and geomorphological development. But the present main features of its relief developed as late as in the neotectonic stage in connection with the formation of the Česká Vysočina (Bohemian Highlands) as an epiplatform orogenic region and the late orogenic stage of development of the West Carpathians. With respect to the complicated relief and the geological structure and the lack of exact data concerning younger deposits the morphostructural analysis appears to be most suitable method of investigating the neotectonic movements and the development of the relief. The morphostructures are understood here as forms developed predominantly by endogenic processes in which both structural-lithological conditions and exogenic processes (mainly in the shaping of passive morphostructures) took part.

In the territory investigated the activized epiplatform orogenic area of the Bohemian Highlands characterized by dense network of faults is the morphostructure of highest order. In this stage of research mainly the area between the Carpathian Foredeep in the east, the deep fault in the central part of the Moldanubian cristalline complex in the west and one of branches of the Labe (Elbe) deep fault in the north was investigated. This area consists of several blocks - morphostructures of lower orders.

The investigations revealed great differences in the character of the morphostructures in territories along deep faults and territories far from these fractures. In the former territories a very complicated tectonic relief of small fault blocks developed with numerous horsts, grabens and tectonic

valleys. The blocks are several hundreds of metres in size
and the amplitudes of movements surpass the value of 400 m.
Most intensive faulting occur on the western and eastern margins of the territory. The central parts are less affected,
the blocks are larger. Here rather structural-lithological
properties of basement, for instance on granites and limestones come into play.

The second territory covering a larger area exhibits a
less disturbed structure. It is characterized by relief forms
designated as fold-faulted morphostructures. These are blocks
of greater dimensions (kilometres and tens kilometres) characterized by a dome-like deformation. A whole spectrum of forms
can be distinguished as the result of combined effects of
young fault and fold tectonics, both the fault and fold components being manifestations of vertical movements. Theoretical-
ly tree groups of transitional forms between block and fold
morphostructures can be distinguished:

anticline	brachyanticline	dome
horst anticline	horst brachyanticline	dome-like horst
anticlinal horst	brachyanticlinal horst	horst-like dome
horst	horst	horst

The fold-faulted morphostructures are characterized by
an irregular groundplan depending on the course of the faults,
by deformed plantation surfaces often with remnants of Tertiary deposits and by a radial (in details rectangular) valley
pattern. The surfaces of planation kept preserved best along
longer axes of fold-faulted morphostructures. On the contrary,
the margins are strongly tectonically disturbed and bordered
by satellite blocks. A striking feature of the fold-faulted
morphostructures is the brachyanticlinal character of their
structure. They manifest themselves first in the fact that the
highest points of the morphostructures are situated excentrical-
ly in the direction towards the Bohemian Highlands, second
that the longer axes of the morphostructures are tilted in
harmony with the general geomorphological conditions towards
the orographical margin of the Bohemian Highlands. An analysis

has shown that the longer axes of all fold-faulted morphostructures run perpendicularly to the margin. From this fact it follows that the origin of the morphostructures is genetically connected with the formation of the big monocline of the eastern margin of the Bohemian Highlands on which these forms repose. This is proved even by our investigations in the Outer Carpathian Depressions where they occur on less resistant Miocene rocks and are therefore strongly denuded. Exceptions can be found in places of outcrops of more resistant sediments such as Lithothamnion limestones on which dome-likely deformed structural surfaces could keep preserved.

Kalinine A.M.
(URSS, Moscou)

SURFACES D'APLANISSEMENT ET PARTICULARITES DES PROCESSUS D'EROSION EN AFRIQUE CENTRALE

Le relief de la République Centrafricaine est dominé dans son ensemble par la monotonie de la plate-forme oubanguienne légèrement ondulée. Cette monotonie du relief n'est rompue que par quelques crêtes de quartzites, dômes de charnockite et inselbergs de granite. A l'Ouest et à l'Est du pays se trouvent cependant deux zones montagneuses qui occupent à peu près 2% du territoire. Les nombreuses rivières drainant les pentes de la plate-forme ont les vallées ouvertes partagées en plusieurs biefs par des rapides formés des seuils rocheux. Les débits généralement varient fortement d'une saison à l'autre.

On distingue du Sud au Nord, en bandes sensiblement parallèles trois zones climatiques nettement différenciées: a)- zone subéquatoriale; b) - zone tropicale; c) - zone sahélienne.

La durée moyenne de la saison des pluies est assez variable et décroît du Sud au Nord. Elle s'étend de 9 mois dans la zone subéquatoriale, 7 mois dans la zone tropicale et 6 mois dans la zone sahélienne. Les précipitations annuelles sont réparties sur 1700 mm au Sud et 800 mm au Nord du pays.

La température annuelle moyenne varie suivant les régions entre 25 et 32°. Presque toute la région étudiée se trouve dans la zone de savane. La forêt dense occupe 5% du territoire.

Les terrains centrafricains se répartissent en deux grands ensembles: les terrains du soubassement plissés, disloqués, métamorphiques d'une part, et d'autre part les terrains de couverture subhorizontale, exclusivement continentaux. Le complexe de base affleure sur 61% de la superficie de la région étudiée. Il peut être subdivisé d'après les données de J.I.MESTRAUD en 6 ensembles lithologiques: des faciès, schistes quartzites, micaschistes, gneiss, amphibolhites et des formations charnockitiques et granitiques de l'âge précambrien D.

La direction générale de la structure tectonique de ce complexe est Nord-Sud.

Le groupe supérieur forme un fragment septentrional de couverture précambrienne A. Ce groupe est représenté par des faciès silicaux, argileux; argilo-silicaux des carbonates, tillites, conglomérats et brèches. Ils ont souvent un aspect tabulaire et non métamorphique. Ces roches se sont formées dans les conditions de la sédimentation épicontinentale rangée au Sud de l'Oubangui.

La série secondaire recouvre 32% de la surface totale. Elle est représentée par des grès et des conglomérats à gros galets roulés, provenant du socle précambrien. De nombreuses lentilles argileuses alternant avec des grès fins et gros ont une stratification entrecroisée, témoignage d'une origine continentale fluvio-lacustre. Les niveaux terminaux pourraient avoir une origine éolienne. L'épaisseur de la série varie de 20 à 300 m.

La formation du continental terminal est formée par les assises gréseuses d'âge paléocène à pliocène supérieur, qui affleurent sur la bordure sud de la cuvette tchadienne. Dans la région étudiée, ces assises sédimentaires reposant en discordance sur le socle sont les suivantes: sables rouges, grès

27

conglomératiques, grès arkosiques, argiles blanches et des niveaux interstratifiés, des latérites.

La formation neo-tchadienne d'âge quaternaire couvre transgressivement au Nord-Est du pays, le socle ou les grès secondaires. On distingue des alluvions récentes occupant le fond de dépressions et des alluvions anciennes, formant les sommets des interfluves.

Les alluvions récentes comportent des argiles grises qui correspondent à des dépôts lacustres et des sables fins du lit de rivières. Les alluvions anciennes sont constituées par des sables plus ou moins argileux. Leurs épaisseurs sont comprises de 1 à 5 mètres.

La formation superficielle actuelle ne se présente pas en tapis unique, mais elle couvre d'une façon irrégulière et discontinue des dépressions et les pentes des interfluves.

La reconstitution paléogéographique se basant sur une interprétation morphogénique nous permet de déduire les principales étapes qui ont déterminé tout le développement du relief récent.

Au début du précambrien supérieur, la région étudiée se transforme dans un bouclier. Cette phase géologique très ancienne a été suivie d'une première période de dénudation et de la formation d'une surface aplanie. La composition des dépôts précambriens supérieurs témoigne des conditions de sédimentations à faible profondeur dans un bassin épicontinental en bordure d'un pays fortement pénéplané. Ces deux éléments morphostructuraux, la pénéplaine et la plate-forme ou la plaine épicontinentale de précambrien ont une large expansion sur tout le continent africain. On peut estimer que le régime tectonique des périodes postérieures dans la région étudiée était celui de plate-forme. Les déformations locales n'apparaissent pas des changements principaux dans le plan structural du pays.

Après une longue période de dénudation et d'érosion au début du secondaire, la pénéplaine précambrienne a été recouverte par des grès continentaux. Ces grès ne sont pas plissés

et sont dépourvus de toute trace de métamorphisme. Leur laté-
ritisation est assez faible.

Au cours du secondaire sur la surface de la pénéplaine
réduite, se forme une plaine fluvio-lacustre avec le relief
désertique.

Les déformations néotectoniques ont provoqué l'effondre-
ment de la cuvette tchadienne et entraîné la formation d'un
réseau de failles et la surélévation de la surface d'érosion
antérieure.

Au paléogène, a donc commencé un cycle d'érosion dont
les dépôts constituent la plaine fluvio-lacustre d'âge paléo-
cène-pliocène.Au pléistocène,après un léger affaissement attes-
té par la différence d'altitude entre les terrasses alluvia-
les, se sont déposés les sables et argiles constituant la
plaine alluviale-quaternaire. Ce nouveau cycle d'érosion a
provoqué un rajeunissement du relief dans les parties émer-
gées. La pénéplaine vient d'être partiellement exhumée de la
couverture du crétacé supérieur,   dégageant ainsi la surface post-hercynienne. Dans la région étudiée, l'évolution du
relief est déterminée par des cycles d'érosion et de dénudation successivement fixés dans les surfaces d'aplanissement.

Selon le schéma proposé le premier cycle dont la durée
est la plus longue se termine en précambrien supérieur par
l'édification de la surface d'aplanissement polygénique, y
compris la pénéplaine et la plaine épicontinentale. Le second
cycle englobe le premier et le secondaire; il s'achève en
crétacé supérieur par l'édification de la surface d'aplanissement polygénique mésozoïque comportant la pénéplaine réduite, la plaine épicontinentale et la plaine fluvio-lacustre. Le troisième cycle commence au paléocène et se manifeste
par une complication ultérieure de la morphologie structurale
du pays. La surface de l'aplanissement polygénique de ce dernier cycle comprend la pénéplaine réduite, la plaine épicontinentale, la plaine fluvio-lacustre du crétacé supérieur, la
plaine fluvio-lacustre paléocène-pliocène, la plaine fluvio-
-lacustre quaternaire et la pénéplaine partiellement exhumée
de la couverture du crétacé supérieur.

Dans le relief actuel les formes de dénudation dominent celles d'accumulation.

La pénéplaine oubanguienne occupe la grande surface des deux bassins: tchadien et oubanguien, séparés par une dorsale faiblement marquée dans le paysage. Les altitudes les plus élevées situées en effet sur cette ligne de partage des eaux sont de 550 à 700 m, décroissant vers l'Oubangui jusqu'à 400 m. La pente moyenne est de l'ordre de 4 pour 1000. La monotonie du relief est troublée par les arêtes de quartzite ou quelques collines de roches cristallines peuvent produire des nivellations assez brusques mais qui n'excèdent jamais de 100 à 200 m maximum. Le réseau hydrographique est relativement dense sur les granites-gneiss et plus lâche sur les micaschistes et quartzites en moyen de 0,22 km/km^2, le coéfficient sinueux de lit - 1,77.

La plaine épicontinentale considérée dans son ensemble constitue une vaste plaine inclinée dont la pente moyenne est d'ordre de 2,1 pour 1000, avec un relief adouci dont l'altitude ne dépasse pas 450 m dans le Nord et s'abaisse à 340 m dans le Sud. Le réseau hydrographique est dense de 0,136 km/km^2, nettement orienté avec le coefficient sinueux de lit 1,66.

La plaine fluvio-lacustre du crétacé supérieur se présente comme de vastes plateaux isolés de Carnot-Berbérati et de Mouka-Ouadda qui sont vallonnés et découpés par des vallées évasées mais parfois ayant l'aspect de véritables canyons. Elle est bordée sur la mageure partie par des falaises abruptes avec le socle précambrien exhumé. Ces ruptures de pentes ont une grande amplitude de 10 à 120 m. Pour l'ensemble, l'altitude s'élève régulièrement de Sud au Nord de 500 m à 800 m. La pente moyenne est d'ordre de 3,5 pour 1000. Les rivières avec ces affluents en drainant des plateaux sont caractérisés par un reseau moins dense de 0,135 km/km^2 avec le coefficient sinueux de lit 1,35.

Le caractère essentiel de la plaine fluvio-lacustre paléocène-pliocène réside dans l'absence presque totale de dénivellations topographiques importantes. L'altitude ne varie

guère de 440 à 390 m, et décroît régulièrement vers le Nord-
-Est. La pente est de 0,6 pour 1000. La morphologie de cette
plaine est caractérisée par de larges interfluves surbaissés
avec pentes faibles et des vallées largement ouvertes. La densité de réseau hydrographique à peine marquée dans la topogra-
phie est de 0,11 km/km^2; coefficient sinueux de lit est de
1,31. Au Nord-Est, elle est replacée par la plaine alluviale
quaternaire à laquelle correspond l'altitude la plus basse
de 410 m en Ouest et de 380 m en Est. La surface est rigoureuse-
ment horizontale, les vallées à peine marquées, leur densité de
reseau est de 0,07 km/km^2 et le coefficient sinueux de lit 1,21.

La recherche des processus d'érosion permet de découvrir
5 zones qui reflètent la correlation des facteurs de modèle
du relief. Ces zónes sont suivantes: 1º - Zone d'une forte
intensité d'érosion aréolaire et d'une faible capacité du
transport des matériaux par courant du lit.

2º - Zone d'une forte intensité d'érosion aréolaire en
association avec l'érosion latérale du lit.

3º - Zone d'une érosion profonde du lit en association
avec une forte intensité d'érosion linéaire des pentes.

4º - Zone d'une érosion aréolaire faible en association
avec une accumulation des matériaux transportés par le cours
d'eau.

5º - Zone d'une stabilité d'érosion aréolaire et d'érosion fluviale.

Kral V.

(ČSSR, Prague)

PIPING PHENOMENA IN BOHEMIA

Piping is a physical mechanical process of downward moving water carrying in suspension fine grain particles of clay
and silt into fissures and percolation paths of groundwater,
or a process of transport of the particles to the outlets at
the foot of slopes. Chemical dissolution may also participate
in the piping process when the clastic rock also contains dis-

31

soluble components, for instance in the rock's cement. Piping
results in saucer-shaped depressions, sinkhole-like or kettle-
like depressions and blind valleys. Underground piping produces tubular (pipe-like) cavities which sometimes constitute
extensive systems of caves, several hundreds metres long. As
phenomena arising is insoluble, i.e. non-karstic rocks are involved, they may be termed as pseudokarstic phenomena.

In Bohemia, piping phenomena and processes are known in
various geomorphological and geological conditions. In North
Bohemia, sinkhole-like depressions are frequent in the loess
covering dissected massive quartzose sandstones (Quadersand-
steine) of Cretaceous age. The sandstones are fissures by a
system of vertical open joints of two main directions; the
sinkhole-like cavities arise by scouring waters washing down
toward underground the overlying rocks, loess and sandy weathering products. Pseudo-sinkholes circular or oval in shape are
most frequently 10 - 15 m in diameter, being 5 - 25 m deep. A
developmental series of such pseudo-sinkholes has been established, ranging from shallow saucer-shaped waterlogged depressions to deep funnel-shaped forms which bottom is fractured
showing open joints in the underlying sandstones. Piping may
also explain the sinkhole-like depressions (pseudo-sinkholes)
in the weathering products of Tertiary volcanic rocks in West
Bohemia. These forms arose on the surface of basalt or tra-
chyte table mountains which represent denudation relics of
lava nappes. The volcanic bodies are dissected by vertical
and subvertical fissures. The washing down of weathering products (loam and debris) into open joints produced longitudinal
sinkhole-like depressions (pseudo-sinkholes) depending on the
strike of the joints. The largest of them are 40 - 50 m long
and 3 - 5 m deep. In West Bohemia, piping phenomena have also
been established in Paleogene sandstones and conglomerates and
their weathering products.

The platform consisting of fresh water Tertiary sandstones of the Sokolov Basin is cut by a narrow, 40 m deep valley of the river Ohře. On the flat surface at the margin of
the rocky sandstone walls, there occur frequent pseudo-sink-

holes 5 - 20 m in width and 3— 5 m in depth. In the sandstone crags standing out from the slopes of the valley, many
rock shelters originated, as well as caves developed from fissures narrowing toward the interior of the rock massif. The
longest cave, horizontal, is more than 40 m long, communicat-
ing with two open pseudo-sinkholes. In this case a relationship between the origin of pseudo-sinkholes and caves is evident; however, not only piping played a role in the origin of
caves but also the lateral erosion of the river during the
respective excavation stage of the stream, in addition to
frost processes and collapse of rocks.

From the above-mentioned examples of piping phenomena in
Bohemia some general knowledge may be inferred, which can be
summarized as follows:

1. In Bohemia, piping phenomena and processes are known
in cohesionless deposits of loess, fluvial sediment and weathering products in the cases where they are underlain by rocks
dissected by a system of fissures enabling removal of fine material toward lower levels or its transport to the outlets at
the foot of slopes.

2. Piping processes in Bohemia occur especially in layers
at the margins of platforms bounded by steep, often rocky
slopes. From the distribution of the phenomena under study it
follows that the relative height of the platforms above the
local erosional base directly influences the intensity of piping processes.

3. The examples of piping phenomena and processes in
Bohemia, given above, actually represent part of slope modelling in their subterranean component.

4. Opening of fissures in the underlying rocks, which is
a prerequisite of the spreading of piping processes, is due,
in Bohemia, to several factors, especially frost action under
periglacial climatic conditions, furthermore, to release of
mountain pressures when the weight of the rocks diminishes by
denudation, and, in some cases, also to landslides.

5. On the basis of the age of the covering formations in
which the piping phenomena of the surface originated, and on

the basis of the morphology of the surrounding landforms it
can be stated that, in Bohemia, the landforms produced by piping are of Holocene or, at most, Late Pleistocene age.

Pérez Hernández A., Rodrígues Otero C.M.
(CUBA, Havana)

MORPHOMETRIC METHODS APPLIED TO THE REGIONAL PLANNING OF THE ISLE OF PINES

This work deals with geographical space differentiation
in territorial units according to morphometric characteristics of the relief, in order to get the best results in agriculture (crops and livestock) as well as for the localization
of different investments.

It is precisely on the relief where most part of the human activity takes place, directly or indirectly influencing
the erosive state of soils, vegetation, the climate, the agriculture practice, the settlement and development of urban
centers, etc., from which arises the need of interpretation
through morphometrical analysis including:

1. Altimetry.
2. Slope angle and direction $tg\alpha = \frac{h}{l} \times 100$.
3. Vertical dissection $\Delta H = H_{max} - H_{min}$.
4. Horizontal dissection $D_h = \frac{Lr}{S}$.
5. Potential fluvial erosion $H_m = \frac{(n-1) \neq (n-1) e}{2}$; $I_{pe} = \frac{Hm \times L}{10^6}$

This study will use one square kilometer as the area unit
and will be drawn on maps (scale of 1:50 000) to be used only
in - those areas where superficial drainage is the basic factor in the modelling of relief.

Values and intervals are used which permit the measuring
of - differing characteristics in each territory using internationally accepted scales and adjusting them to the results
obtained in the investigation.

In Isle of Pines, where the most significant FEATURES are
a - Denudational relief; b - abundance of low and marshy zones

c - northern sector with superficial drainage network on kao-
linitic and ferrolitic weathering crusts and a meridional
sector of subterranean drainage, characteristic of intense
karst development zones. The main morphometric characteristics
are as follows:

1 - Altimetry intervals of 0-5, 5-10, 10-20, 20-40, 40-60,
60-80, 80-100 > 100 expressed in meters above sea level
(91.58% of the area is - below 40 meters, while only 0.5% is
over 100 m).

2 - Steepness angle and direction: 0-0.5% (very low
plains); 0.5-1% (horizontal plains); 1-2% (light slope);
2-5% (slightly inclined); 20-25% (scarped) and 25% (strongly
scraped).

3 - Vertical dissection: the following ranges were considered: 0-2.5 m (very low and very slightly dissected plains);
2.5-5 m (transition); 5-10 and 10-20 m (slightly dissected
plains); 20-40 m (moderately dissected plains); 40-60, 60-80,
80-100, 100-120 > 120 m (very dissected heights , with concentric characters and limited - extensions).

4 - Horizontal dissection in the northern section:
0-0.3 km/km^2 (very slightly dissected), 0.3-1.km/km^2 (slightly dissected); 1-2 km/km^2 (moderately dissected); 2-3 km/km^2
(very dissected); > 3 km/km^2 (extremely dissected).

5 - Potential fluvial erosion: where drainage density
and medium-height per km^2 are combined, the selected ranges
are: 0.000-0.005 (very slight to slight); 0.005-0.02 (moderate); 0.02-0.1 (severe) > 0.1 (very severe).

Once the mapping is finished and evaluation of the territories included in each category is separately undertaken
within the framework of the analysis and of aerial photos and
field checking: the next stage deals with the identification
of coincident categories which may be superposed:

a - Slope and vertical dissection maps, previously determining restrictive ranges (by excess or by defect) for the use
of areas for agriculture and cattle raising, building and so
forth, serving as a mean for checking the map of permanent and
periodical flooding of the territory.

35

b - The maps of horizontal dissection and potential fluvial erosion, showing a particular trait of the territory and allowing for the establishment of areas (fit or with limitations) for the basic uses.

A due superposition from both partial results allows for the establishment of category or relief according to the above mentioned morphometric methods. Adopting for the following restrictions:

Territories	Unfit	Fit with limitation	Fit	Fit with limitation	Unfit
Slope (%)	< 0.5	< 0.5	0.5–10	–	> 10
ΔH (m)	< 2.5	< 2.5	2.5–20	–	> 20
Dh (km/km)	–	–	0–2	2–3	> 3
Ipe	–	–	0.005–0.02	0.02–0.1	> 0.1
Observations:	Marsh	Seasonal flood	Without limitation	Erosion	Accelerated erosion

A total of 5 categories were obtained:

CATEGORY I. Very low plains slightly dissect and marshy lands - (2 m) not useful in their natural state.

CATEGORY II. Very low plains slightly dissect (5 m) of restricted use because of periodic flooding requiring drainage works for its complete incorporation to agriculture and cattle raising. Not recommended for building purposes.

CATEGORY III. Very low plains moderately dissect (5 m to 40 m). They constitute the most favourable for agriculture and cattle raising. Allowing for total mechanization and irrigation design of typical fields for cultivation and so forth. They are useful for - building for soil movements and cost diminish, the possibility to flooding is reduced to some sectors when hurricanes pass.

CATEGORY IV. Dissected high plain (40 - 60 m); agricultural use restricted by high density of drainage and erosion, requiring the application of antierosive techniques, raising of perennial crops, reforestation, which results in useful of benefits of the water reservoir life although the net area diminishes.

CATEGORY V. Heights very dissected (60 m), recommended only for the preservation of the ecosystem, touristic potential exploitation and reforestation.

The karst meridional plain, characterized by skeletal soils, the presence of lapiaz, casimbas, skin holes and so forth is not to be immediate agricultural use and it is recommended for the ecosystem preservation and forest exploitation.

Zapletal L.
(ČSSR, Olomouc)

A NEW METHOD OF THE GEOMORPHOLOGICAL CHARACTERIZATION OF AN ANTHROPOGENIC RELIEF

The XIX Geographical Congress in Stockholm, 1960 drew the attention to the fact, that in the contemporaneous century of the scientific-technical revolution in economically developed countries with the well developed culture the man presents a most energetic and most effective external geomorphological factor. In some countries the rapid development of the anthropogenic geomorphology has happened since, that led to a proposal of the genetic-morphological classification of the anthropogenic forms of the relief, to the stabilization of the normativa and the terminology of the new scientific branch and to the distinguishing the anthropogenic geomorphological processes direct from indirect and the tertiary ones inclusive antiprocesses. After the stadium of characterization the separate anthropogenic relief forms (in number of which is presently distinguished 63 species different morphologically and genetically, belonging to 9 different groups of the genetic-morphological system) comes stadium of the complex characterization of anthropogenic relief as a whole. To this geographical characteristic this study presents a proposal of the new method.

The quantitative characterization of the set of the separate anthropogenic relief forms on the earth surface demanded data only of the number of the single species and types of the anthropogenic relief forms in the open air completed by data

of largeness of these forms of the earth surface. Because of
considerable difference of largeness, form and basic difference of the anthropogenic relief forms of the single territories these data were for various territories mutually uncom-
mensurable. The demand arose to characterize the set of the
anthropogenic relief forms complexly as an anthropogenic relief.
The classical characteristic applied through a long time expresses the area of set of the anthropogenic relief forms in
the given territory unit. This characteristic made possible
indeed comparison of the separate areas from point of view of
the anthropogenic geomorphology, but it was not objective,
because the anthropogenic relief has on the different places
the various and as a rule considerably different thickness.
Therefore we propose as a new method of the geomorphological
characterization of the anthropogenic relief s.c. anthropogenic geomorphological effect.

The anthropogenic geomorphological effect is a numerical
datum of thickness of an abstract stratum that would come into
being through the equal spreading the matters transported by
anthropogenic agents of the soils on the whole examined territory. The construction of the map of the anthropogenic relief is established upon the theory of the cartographic utilization of small territories, for which the anthropogenic geomorphological effect is calculated as a proportion of the summary volume of the matters transported through anthropogenic
aggradation, degradation, planation and excavation and of the
area of territory according to the following formula:

$$e = \frac{\sum_{i=1}^{9} A_i}{P}$$

in which e is the value of the anthropogenic geomorphological
effect given in metres, P is area of the territory given in
m^2 and $A_1 - A_9$ is the volume quantity of soils given in m^3.
The volume of the transported matters $\int A \int$ being given in
cubic units and the area of territory P is given in quadratic
units, has e length value. The anthropogenic geomorphological
effect does not characterize terrain from the point of view of
the form not even of the genesis of anthropogenic forms, but

it characterizes complexly quantity of the transported soil on the single places. As a rule it is given by the value for a territory in extent of 1 km^2.

The substitution of the values found cartometrically or in the terrain in the present formula is arithmetically very easy but in a large number of the anthropogenic relief forms in the terrain – very difficult. Application of the mechanical technique, of the space differentiation of territories of the various type through the computers proves here very well. In a duly settled program and $\underline{P}$ being for all territory units equal, the computer settles directly the values of the anthropogenic geomorphological effect. But even without application of the computers it is possible to realise the calculations. We have experimentally constructed a map of the anthropogenic relief for territory divided by us into 23 874 squares in extent of 5.29 km^2 and for every of these territories we have worked out the value $\underline{e}$.

The geographical applicability of geomorphological characteristic of the anthropogenic geomorphological effect was examined recently by some authors namely for territories of the various extent from ten km^2 to hundred thousands km^2. Characteristics of the anthropogenic relief according to the values $\underline{e}$ were published for Czechoslovakia (127,870 km^2), Slovakia (49,008 km^2), for North Moravian region in ČSSR (11,066 km^2), for the promontory landscape of Zailiski Alatau (Middle Asia, 5,800 km^2), district Olomouc in ČSSR (1,449 km^2) and for territory of the town Krnov in ČSSR (52 km^2).

The cartogram of the calculated numerical values $\underline{e}$ makes possible in a simple arithmetical way to settle (as a proportion of sum of the values $\underline{e}$ in all parts of the given territory and of the number of the parts in that territory) coefficients of the anthropogenic geomorphological effect for determined parts of the examined territory: for arbitrary territory physical-geographical, economic-geographical or administrative, f.i. for a village, district, region, orographic whole, economic region et cetera.

The demand of the proposed here geomorphological characteristic was asked by the modern heightened number of treatises and studies on anthropogenic relief in the geographic milieu after an information, that anthropogenic transformations of relief in a dynamical milieu of life are irreversible unlike the anthropogenic changes in hydrosphere and atmosphere for the given places of anthropogenic degradation. The consistent study of the phenomenon also led to settling regards of the values of the anthropogenic geomorphological effect to the other geographical factors. Examining the relation of e to the basic phenomena was f.e. settled the correlation with the altitude of territory above sea-level. There also were fixed more complicated dependences f.e. correlation between measure of anthropogenic geomorphological effect and density, activity and common quality of population of a country. An argument for the existence of these relations they are not only correlation graphs, but also regionalization realised for various territories by the methods of anthropogenic geomorphology.

The genetic-morphological classification of the anthropogenic form of relief proposed by us is briefly published in English in the materials International Geographical Union – Commission on Geomorphological Survey and Mapping (Manual of Detailed Geomorphological Mapping, 1973, pp. 242 – 243).

Bentley G.R., Robertson J.D.
(U.S.A., Madison, Wisconsin)

GRAVITATIONAL EVIDENCE FOR RETREAT OF THE EASTERN MARGIN OF
THE ROSS ICE SHELF, ANTARCTICA

As part of a project to study the dynamics and history
of the Ross Ice Shelf, Antarctica (RISP), a survey of gravity
measurements over the eastern half of the ice shelf has been
completed. Ninety-one stations, basically on a 55 km grid system, were occupied; elevation measurements and seismic reflection determinations of water depth were included. Isostatic
gravity anomalies calculated from these data show a significantly negative mean of -10 ± 1 mgals, with a trend from about
-5 mgals in the center of the shelf to about -15 mgals near
the boundary between grounded and floating ice. We interpret
this pattern to reflect incomplete isostatic rebound following
late Quaternary retreat of the grounded ice sheet. Numerical
modelling suggests, however, that, contrary to the implications
of glacial geological studies by Denton and others in the
McMurdo Sound region, the time of retreat was more than 6000
years ago.

Boytel F.
(CUBA, Santiago de Cuba, Oriente)

ORIENTE'S PROVINCE EOLIC CHART

In this paper are described the technical resources that
had been employed for surveying the first eolic chart in
Oriente's Province, which covers an area of 35,000 square kilometers. Its history brings out the scarce information exist-

ing up to 1960, and urgent needs for eolic information in order to attain solutions in: town and country planning, planning of industril development's areas as well as sanitary and agricultural studies.

Basic information used was obtained from four heterogeneous meteorological stations, only one of them reaching a 49 years uninterrupted period.

The exploration and trial methods have been used to obtain the wind's persistence and its force relationship, which provokes vegetation's eolic deformation.

As such a deformation exists, in a fitted, secular and continuous manner within the whole studied area, the so-called flag effect was carefully analyzed, proposing a regional classification that has been applied, obtaining a chart in a scale 1:250 000, sufficiently precise for the desired objectives. In this way, a first reasonable approximation of the wind's direction, force and frequency is obtained.

For a better analysis of the territory a typology of winds has been obtained, bounding with approximation lines, areas which correspond to different types of winds, including: trade-winds marines breezes, land breeze, gravitational and local winds such as catabatics, and so forth; it also represented the draw of the imaginary line that followed the axis of each hurricane occurred within the last hundred years.

The author has established a number of graphics of beaufort type winds, with modifications which represent the results of field investigative work and of deductive analysis, under the condition of modifying them when the information from the dense network that has been set up, so deserves it.

Conclusions:

It has been possible to make a survey with scarce information from conventional instruments, using the trace left by the wind, -eolic tracing according to the author-in the vegetal mantle. Throughout previous years, this chart has proved to be sufficiently approximated, for its application on physical planning.

It is explained how the author-based on the "Eolic Chart" could establish the validity through a test conveniently approximated, of data stated in the chart, to be used in agriculture and sugar industry in improving cultivation's techniques. The work covered less than a third of the province's surface, and was carried on by team of researchers, with the help of conventional agricultural aviation. A total daily work yield of approximately 1000 square kilometers was done, giving as a result a demonstration of the application of this method in a developing country.

Fernandez Veiga N.
(CUBA, Havana)

EVALUATION OF THE PLUVIOMETRIC NETWORK IN THE NORTHWEST REGION
OF ORIENTE PROVINCE, CUBA

The urgent need for obtaining efficient pluviometric data to meet the purposes and necéssities of our development with regard to the optimum use of the hydraulic resources of our country, requires an objective evaluation of the pluviometric network present state.

There is not an internationally accepted idea of network, but in Hydrology this subject has reached a remarkable development in the last years. The hydrological network design can be accepted like a continuously changing process depending on the hydroeconomic conditions of each region.

. The statistical-meteorological analysis to find the station rational density consists of securing the accuracy degree which the wanted information is to be obtained in this work, for example, the annual rainfall totals, and by means of that analysis to determine the admissible maximum distance among the rainguages. These purposes can be attained by the study of statistical structure of the precipitation field. This study is carried out in Cuba for the first time in order to determine the optimum rational density of the pluviometric network in the Northwest region of Oriente Province. This region has an extension of 4200 km^2.

To carry out the through investigation of the statistical structure of the precipitation field, sufficiently long homogeneous series are needed. The data of 35 raingauges, with 50 years of observation (1924 - 1973), located in the study area, were used for this work.

If we take into consideration the point pluviometric series, spatially separated by the distance $\underline{L}$, we can establish a correlation coefficient $\underline{r}$ ($\underline{L}$) between two pluviometric series, and the independent variable will be the distance L between the two points. The function has to fulfil the condition of homogeneity and isotropy. That is, this function does

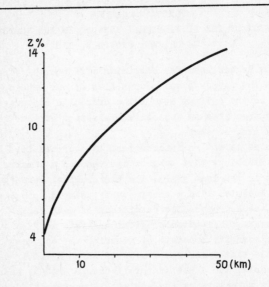

The error function of lineal interpolation for a triangular pluviometric distribution (in the study area)

not depend on the direction of L but only on its distance. According to Kagan (1972) the correlation function is described satisfactorily by means of the exponential relationship such as:

$$r(L) = r(0)e^{-L/L_0} \qquad (I)$$

Where:

L   – distance between pairs of stations;
L_0 – typical radius of correlation;
r(L) – correlation coefficient for the distance L;
r(0) – correlation coefficient for the distance L = 0.

In order to evaluate (I) in the study area, 595 correlations are used (5), the minimum quadratic exponential relationship founded is:

$$r(L) = 0.93 \; e^{-L/71} \qquad (II)$$

To evaluate the interpolation maximum error, in annual average percentage, taking into consideration the pluviometric network which is spatially distributed in triangular form (3) the following is used:

$$Z = C_v \sqrt{\frac{1}{3} + r(0) - 2r(-\frac{L}{3}) + \frac{2}{3} r(L)} \qquad (III)$$

That is to say, the raingauges are located in the vertexes of imaginary equilateral triangles with side L and the relative maximum error of interpolation $\underline{Z}$, is located in the points where the three medians were cut.

The error function was traced using a coefficient of variation C_v for the annual rainfall of 0.25.

Of course, the determinations of the admissible distance among the raingauges with an optimum interpolation error obtained by the error function, have to be considered only like the first step to solve this problem.

Another very important step during the planning and organization of the pluviometric network, is the correct election of the station and its representativeness in relation to the macro-scale conditions.

Hanuška L.
(ČSSR, Most)

POLLUTION ET DEPURATION AU TERRAIN INDUSTRIEL

Le Bassin du Charbon brun de la N-O Bohême appartient
aux plus significatives agglomérations de la ČSSR. Le dépar-
tement de la Bohême du Nord, un des plus petits, fournit au
réseau de la République plus de 35% de l'électricité et ex-
trait 72% du charbon brun. L'industrialisation intensive influence considérablement la nature, l'environnement. De là
proviennent la dégradation du sol, la contamination des eaux
et de l'air.

Par exp. la rivière Bílina (ancien nom Běla = Blanche),
l'axe du Bassin, est l'une des plus polluées en ČSSR.

L'amélioration partielle (station Chánov) nous pouvons
constater après la construction de la station de dépuration
à l'usine de l'organochimie. La dévastation du sol s'était
produite par l'extraction superficielle du charbon et les
terrils. La dégradation suivante de la qualité est conditionnée par désséchement, après l'interruption des loges aquaducs
et par les procès physico-chimiques.

Les usines industrielles émettent dans l'atmosphère
presque 180 000 t/année du dioxyde de soufre (SO_2) et
160 000 t/année du cendre. SO_2 après l'hydratation par l'humidité atmosphérique augmente l'acidité des précipitations
(pH plus bas) et de telle manière aussi du pH des eaux super-
ficielles (et souterraines) et du sol. Dans le cendre du
charbon brun du Bassin (aussi que dans les émissions - dans
les particules solides) nous avons constaté 35 éléments de
métaux ou bien leurs oxydes. A 20 km de la source sont étendues les éxhalations, dans les montagnes, sur la surface de
10 ha sont tombés jusqu'à 35 kg des éléments, et de cette
quantité presque 2 kg toxiques (CO, Cr, Cu, Ni, Pb...). Cette
combinaison nous avons constaté dans les précipitations de
145 mm/m^2 pendant le juillet 1971.

Aussi les substances de phénol et beaucoup des combinaisons de soufre, de gaz carbonique et de nitrogène, s'ajoutant aux particules d'aérosol, changent la qualité de précipitations et contaminent l'eau et le sol.

Dans le cadre de notre plan de travail nous avons étudié la possibilité de rincer (dépurer) l'atmosphère par les précipitations (de pluie, de neige) comme la forme principale de la décontamination de l'air. Les exhalations industrielles et communales changent considérablement non seulement la chimie de l'air, mais aussi les conditions physiques (de la chaleur, de la pénétration de lumière, de l'albedo, etc). On constate l'inhibition des processus physiologiques des plantes (ainsi que des animaux et de l'homme).

Le dioxyde de soufre, plus lourd, bloque le dioxyde carbonique utilisé pour l'assimilation; les métaux conditionnent de l'oligodynamie jusqu'à la toxicité; les acides (H_2SO_3, H_2SO_4) gâtent la pellicule et préparent les voies pour les infections.

Nous avons démontré que l'acidité des précipitations dans les Monts Métalliques (Krušné hory) va de la source des exhalations aux montagnes. A la station HA, pour un réservoir de l'eau potable, dans une vallée éloignée de la source des émissions de 15 km, la concentration de SO_2 était dans les limites de 0,00 jusqu'au 0,08 mg/m^3 en 63% des exemples, mais la concentration des ions d'hydrogène en 74% des exemples atteint pas plus que 5,0 pH. A la station KO, près de la source d'émission du SO_2, le volume de ce gaz dans l'air était jusqu'au 0,08 mg/m^3 seulement en 27% des cas; de 0,08 à 0,35 mg/m^3 déjà en 67% des exemples. La réaction du pH, au contraire, était plus bas (5,5-6,0). C'est un phénomène paradoxal. Et c'est pourquoi on peut constater que le SO_2 ne peut pas seulement être le donateur de l'acidité des précipitations. Cela affirme aussi la théorie de l'intoxication inverse.

Nous avons tentés à l'explication de ce phénomène en connexité de la masse (mg/l) et de la grandeur des particules

solides. Dans l'agglomération industrielle, approchée de la
source des émissions, monte l'effet d'alcalinité par des particules solides plus grandes, quand même la concentration du
SO_2 est plus haute. Les petites particules, jusqu'à 5 mm,
possèdent une possibilité de l'absorbtion plus intensive,
elles se trouvent dans l'air le temps plus long et de telle
façon conditionnent - sur les distances éloignées de la
source des exhalations - la baisse du pH.

La relation des particules solides à la station HA avec
les mêmes de la station KO, était établie d'après la masse
(mg/l). La relation des particules solides (la masse mg/l en
% des cas) entre HA et KO présente les données: jusqu'à
50 mg/l:HA - 64,5%, KO - 39%; plus de 50 mg/l:HA - 35,5%,
KO - 61%. Au fond de cette constatation nous présentons la
théorie d'aérosol: "Au terrain éloigné de la source des émissions, où la concentration du dioxyde de soufre libre dans
l'air n'est pas si haute, le SO_2 est significativement lié
aux petites particules lesquelles avec l'humidité atmosphérique forment l'acidité des précipitations. Dans tels endroits
le donateur principal de l'acidité de pluie (de neige) sont
les particules fines, plus petites que 5 mm du diamètre,
mais avec la possibilité d'une grande dispersion".

Inyang, P. E. B.
(NIGERIA, Calabar)

AIR POLLUTION IN CALABAR: A PRELIMINARY STUDY

Air Pollution is becoming a feature of modern cities especially the industrialized ones. Calabar is fast acquiring
the characteristics of such a city. As a result air pollution
is becoming one of its features. The topography of the city,
among other things, has considerable influence on its air
movement. First the buildings exert a roughening influence resulting in air turbulence. This is aided on by the channelling
effects of the "street gorges". The roughness of the surface
is also accentuated by its undulating nature and the existence

of a swamp forest to the south. As a result wind speed is slow-
ed down and diffusion or dissipation of pollutant particles
takes a longer time than otherwise.

The south-west of the city is cliffy and rises to a
height of more than 200 feet. The prevailing winds from the
south-west, when fully charged with moisture, may discharge
its load before diverging and subsiding. These three processes have a warming effect on the city. At night the winds are
usually off shore as the city becomes cooler than the water
around. Unless there is an outbreak of storm the general experience is one of calm. The pollutants let loose from all
sources into the air begin to settle down because of a temporary inversion. This is usually accompanied by dew formation,
misty or foggy conditions in the morning especially during
the dry months. The solid wastes dumped into the many erosion
scars of the city rot and emit very offensive gases into the
air. These are increased day and night by smoke from burnt
wood from thousands of homes in the city. More pollutants released into the atmosphere come from the several small industries scattered over the city.

The most important source of air pollution in the city
so far, is the Calcemco Cement Factory. Its giant chimney,
rising about one hundred feet above the ground, emits pollutants in different forms into the air. The smell of the che-
micals used in preparing the cement can be traced in any part
of the town. The smoke column, which sometimes merges beauti-
fully with the cumulus cloud, can be followed across the city
for several miles away.

A close study of the movement of the smoke column for
fifteen days (between 20th Nov. and 15th Dec.) reveals the
following characteristics:

(a) From about 8 a.m. to late afternoon, the loop of the
smoke varied from near horizontal through vertical to looping
towards the horizontal.

(b) From late afternoon to early night the loop fluc-
tuated between near-horizontal and horizontal. From this time
towards midnight and dawn there was a general decline in
height towards the horizontal which level was usually attain-

ed between midnight and dawn. At sunrise it began again to fluctuate between the horizontal and the vertical.

Winds over the city in November are almost evenly distributed in direction - being south, south-west, west, north-west, north or north-east. The frequency of wind direction from the east and south-east is less than for any of the other directions. The percentage of calms during this month is as high as 13.2%. In December the situation is the same except that there are a few strong winds from the north-west of the city. The percentage occurrence of calms is 11.8%. It is clear from the picture above, that the smoke plume is more or less omni-directional.

Three main factors - the normal temperature lapse rate, earth and solar radiation and wind direction - may explain the behaviour of the smoke. At 8 a.m. the sun is high above the horizon. Its rays begin to warm up the surface of the town effectively; and normal lapse rate of temperature with height is assured. The smoke column being warmer than the air aloft tends to rise towards the vertical position. At mid-afternoon, the period of maximum earth radiation, the decrease of temperature with height was fully established and the smoke plume was at the time practically vertical. In late afternoon the surface of the city began to cool off and the temperature began to increase with height and the smoke column sought the horizontal. There were fluctuations in the general decline which could be attributed to local factors both atmospheric and topographic.

At night the smoke plume showed a steady decline from the vertical towards the horizontal which level was usually attained between 9 p.m. and the eartly hours of the morning. At the horizontal position of the smoke plume an inversion of air temperature lasting a few hours had occurred. The stack is such that its top sometimes reaches above the surface inversion; so the smoke settled at the level of inversion, while the air near the ground remained clear. However, it was usual during the observation to find the smoke fanning away for several miles and many times settling down on the city in the

form of fog (smog) which later was cleared up through the
formation of dew and the deposition of pollutants. It is at
such times that air pollution is at a maximum in the city.

The height of the sources of pollution is important in
the distribution of pollutants. When a tall stack spreads the
pollutants far and wide, the layer of concentration is thin,
and the chances of cleaning up the atmosphere are great. The
Calcemco chimney stack is designed for that purpose. However,
this objective is never fully achieved, firstly because the
larger and heavier particles of soot and dust tend to fall
near the stack. Secondly, some of the particles combine with
condensation nuclei and water droplets and reach the ground before their expected time. Others are themselves hygroscopic
and so precipitate condensation in the immediate neighbour-
hood. Furthermore particles drifting away from the stack may
be caught up in rain falling from clouds at higher levels. It
is interesting to note that during a month's (September, 1975)
observation there were at least five occurrences of rain
showers in the vicinity of the factory more than in any other
section of the city. Also after every rainfall at night, the
factory when in operation, covered most of the city with its
smoke and visibility became poor. Thus the narrow streets,
flanked by packed buildings, become basins of pollutants.

On some days at day time, anticyclonic conditions prevail for a few hours over the city and temperature inversion
may occur a little above the chimney level, so that pollutants
in the warmer air aloft sink to the interface with the cooler
air below. Meanwhile those released in the morning hours and
risen to the same level of concentration although a good quantity still remained suspensed close to the ground. By early
evening cooling starts; but the Calabar countryside gets cooler than the city which meanwhile continues to emit warm air.
A kind of "air draining" starts from the countryside to the
city until between 10 p.m. and 2 a.m. anticyclonic conditions
are re-established. This also is a period of intense deposition of pollutants over the city as it becomes quite smoky
especially when the factory is still at work. Usually the day

time anticyclonic conditions are never established if the city's temperature remains at par with or higher than that of its countryside.

From the brief analysis above it can be seen that air pollution can become a real problem in the city of Calabar if many giant industries are built without much regard to wind direction and human settlement. Future development must take these factors into consideration. The present master plan of the city seems to take cognizance of this in respect of the main city, but little regard is paid to the effects of factories on the suburbs and adjoining villages.

Lewis J.E., Outcalt S.I.
(USA, Michigan)
VERIFYING AN URBAN SURFACE CLIMATE SIMULATION MODEL

The development of any research program is largely dependent upon the questions to be answered by the proposed work. In urban climatology, we believe that a central question is how does land-use interact with weather to modify the urban climate. The mean surface temperature in any area of urban terrain is a unique response to local weather and the physical properties of the earth's surface and substrate.

The nature of the urban surface controls both the magnitude and partitioning of energy at the earth's interface. The site properties which affect the local climate are the thermal properties of the surface (thermal diffusivity and conductivity), surface aerodynamic roughness, surface reflectivity (albedo), and the distribution of moisture (evaporation and transpiration) at the interface. The combination of these surface properties will then establish a mosaic of microclimates in the city, each having its particular radiation-energy budget. Therefore, when the surface properties are modified the phase and amplitude relationships between the components of the surface energy are perturbed.

Present modeling technology permits the numerical simulation of the surface thermal and energy balance regime as a function of local meteorological observations and the radiative, aerodynamic and thermal properties of the near surface environment.

This paper describes the verification attempts of the authors in testing an equilibrium surface temperature simulation model. "Verification testing" is accomplished through the use of aircraft multispectral scanning information which is employed to construct observed maps of surface radiant temperatures for an urban area. The general approach in modeling urban temperature fields is based on the simulation of a set of temperatures associated with surface characteristics representative of a particular land use. The rationale of this method is that each land-use type with its individual mix of surface properties generates an energy regime unique to that type.

Observed Data Source: Multispectral scanner imagery was collected for the city of Baltimore, Maryland on May 11, 1972 at 615, 1015, and 1345 EDT. The aircraft flew at 1524 meters above sea level for which data was collected in the following radiation band $9.8 - 11.7\mu$. Very detailed procedures for calibrating the thermal imagery were followed. The interested reader is referred to Lewis et al. (1975) and Pease (1974). One flightline and time of day was chosen for use in verifying the simulated results.

A smoothed digitized computer map was constructed employing the Stefan-Boltzmann relationship assuming an emissivity of .95. This emissivity assumption is reasonable because the city does not have widely varying types of moisture sources due to the existence of mostly man-made impervious materials, and the range of radiant temperature values for emissivities from .90 to .95 is subsumed in the class interval of 5°C used on the map.

Simulation Model: The details of the simulation strategies are available in the current literature (Outcalt, 1972) and will not be discussed here in detail.

Briefly, the method hinges on the specification of all the variables needed to calculate the components of surface energy transfer based on the meteorological (M) and geographical (G) data. Then, the four components of surface energy transfer (net radiation (R), soil (s), sensible (H) and latent (l) heat flux) can be specified as transcendental in surface temperature (T) in the familiar energy conservation equation.

Eq. 1. $R(G,M,T) + S(G,T) + H(G,M,T) + L(G,M,T) = 0.0$

A suitable numerical algorithm (interval-having or secant) is selected to carry out a search for that surface temperature which will drive Equation 1 to a zero sum condition. Then, an explicit or implicit finite difference algorithm is employed to generate an update soil temperature/depth profile. Thus, at each iteration, the surface temperature and all the components of surface energy transfer are output in addition to the substrate (soil) thermal profile.

The next state of model development required the estimation of the geographical terrain parameters from easily measurable characteristics of urban terrain in sample tracts several blocks in area. It was discovered that the geographical parameters were related to three easily measured features of urban terrain which could be abstracted from stereo air photography, multispectral imagery, insurance maps, etc. The relationships were derived from consideration based on the earlier work of Lettau (1969) and Myrup (1969).

The modeled temperature field is constructed in the following manner. A specific set of geographical (urban) terrain factors (surface roughness, substrate diffusivity, surface albedo, percent of transpiring surface-wet fraction, and silhouette ratio) are estivated from the particular land use. Once the temperatures for the respective land use are simulated, a 40 X 15 grid (one grid square equals .9 km^2) is overlain on the land-use map of the study area and a temperature value is plotted for the land-use type occurring at that grid intersection.

For this study the values of the geographical terrain parameters are obtained using techniques outlined by Jenner

(1975) with the exception of surface albedo and substrate diffusivity. Surface albedo values for each land-use type were as-
sessed from observed multispectral data. Diffusivity values
were obtained from Myrup and Morgan (1972). An internal change
was made in the surface budget simulation model in which sky
radiant temperature is usually set at 22°C below mean air temperature, however for this simulation exercise an observed
mean sky radiant temperature of -5°C was used.

The U.S. Geological Survey's land-use classification
(Anderson, 1972) employed was modified slightly. The residential class was subdivided into low, medium, and high densities
and commercial land use was defined CBD for a specific area
in downtown Baltimore. In addition, park and open category
were combined into one class while institutional land-use temperature was calculated by weighting commercial and open land
use (park) temperature values on a 1/3 and 2/3 bases respectively.

The simulated temperatures were then mapped based on the
grid method previously discussed. Given this grid system 600
temperatures are used to produce the simulated map. The geographical limits of the map are reduced compared to the
observed map. Following a NW-SE direction, the
simulated map begins at the northern end of Druid Hill Park,
continues into the high density residential area on Baltimore's west side, on through the CBD and ends on the south
side on Patterson Park.

The simulated, as well as the observed, map has 7 class
values in which the range of simulated temperatures is from
15°C to 40.3°C. The grey tone shading follows the same tone
progression as the observed thermal map where the darker tones
are the cooler temperature and as the tone decreases, the temperature increases.

The relative spatial association of the simulated vs.
observed thermal pattern is good. Druid Hill Park and Lake
have the cooler temperature with a SE progression into the
much warmer temperatures of the high density residential areas.
Cool islands of vegetation poke through the dominant pattern

of higher temperatures in the form of Greenmount Cemetery and Patterson Park. Also two cooler arms are seen on the west side of the map which are produced by the cooler CBD and the intrusion of the Chesapeake Bay which has the coolest temperature for the entire location. In light of the experiment, we feel the simulation model performs an adequate job of replicating the observed surface radiant temperatures at 1345 EDT. Further analysis is being conducted for other times of the flight.

In an absolute sense, however, the simulation consistently underpredicted the observed temperatures for all land-use types. However, the underprediction is a result of the overall conservative nature of the simulation model and also because the atmospheric damping depth remains constant for a 24 hour period simulation. This constant damping depth produces a steeper than usual lapse rate during the afternoon hours resulting in much greater amounts of sensible energy being removed from the surface than otherwise would occur. It follows that modeled surface temperatures would be underpredicted.

A comparison of the predicted range $25^{\circ}C$ ($15.0 - 40.3^{\circ}C$) versus the observed of $28^{\circ}C$ ($20.0^{\circ}C - 48.0^{\circ}C$), and appropriate hierarchical ranking are considered good results for a relatively untuned simulation model.

Conclusion: This type of analysis may eventually explain the somewhat confused mass of empirical data on the seasonal and diurnal variations in the heat island effect. Further, the work strongly suggests that the physical basis of the "heat island effect" is both geographically and temporally variable, and that an explanation suitable for London's winter may not be suitable for New Orleans or Tucson or even London in summer. Thus, this primitive model indicates the danger of subjecting the phenomena to explanation by "universal hypotheses" as the energy budget equation as developed here has a response sensitivity to geographical parameters variability which is dependent on both weather and the values of other geographical variables. Thus, the system sensitivity to, say, increased

park area (wetness) is not temporally or spatially stationary,
but depends both on the values of obstruction height, silhouette ratio and local weather.

This work on Baltimore points the way toward the development of a new branch of climatology, "terrain climatology",
in which modeling and remote sensing data acquisition operat-
ing at compatible spatial scales can be employed to gain a
sound theoretically-based understanding of the physical basis
of surface thermal contrast.

Lovász Gy.
(HUNGARY, Pécs)
THE VERTICAL ARRANGEMENT OF THE SPECIFIC RUNOFF ($1/s.km^2$)
IN CENTRAL AND SOUTH-EUROPEAN OROGRAPHIC SYSTEMS

The examinations had two purposes:
- The analysis and determination of the vertical arrangement of the specific runoff in Central and South-European orographic systems belonging to different climatic regions. We
tried to explore the connection between different atmospheric
motion zones and the regional arrangement of vertical systems.
- We wanted to examine how the features of the ground in
high mountains influenced vertical values. The role that big
valleys and straits in orographic systems has in the formation
of the vertical arrangement of the specific runoff was also
examined.

The research work was carried out in the Alps, Carpathian
and Dinaric territories, which belong to the Danube river-system. The vertical system consists of the specific runoff values determined at 500, 750, 1000, 1250, 1750 and 2000 meters
above sea level.

The fall or rise in value is shown by two simple parameters:
- The first parameter is a vertical gradient belonging
to a specific runoff value, showing the $1/s.km^2$ value cor-
rects to 100 m above sea level.

Represented: $\dfrac{1/\text{s.km}^2}{100 \text{ m}}$

- The second parameter is a quotient showing the change of the specific runoff. This quotient indicates the proportion of the values of a vertical geological profile at a certain altitude to the values of the lowest altitude (e.g. 500 m).

- In the course of the research work the following genetic types could be determined in the system of the specific runoff:

1. The situation on the slopes of orographic systems facing the prevailing atmospheric motion. Among all examined types the greatest absolute values were found here at each altitudes. Another special feature is the greatest rate of rise in values according to the vertical. We had the highest vertical gradient of the specific runoff here. The Northern slopes of the Alps belong to this type. This territory rises like a huge wall against the atmospheric motion coming from the North, North-West. The rise of air masses and the formation of precipitation caused by orography is the most intensive in this area.

2. The second type is represented by the inner territories or the Alps. These lie near to the Northern ridge, and are broken up by longitudinal valleys (Inn, Salzach, Enns). Lower values of the specific runoff ($1/\text{s.km}^2$) are characteristic of this area at altitudes mentioned earlier. The vertical gradient of the specific runoff is significantly less as well. From 750 - 1000 m above sea level it is zero or with negative sign. In this type orographic rains play a less important role. This explains the smaller values of the two parameters of the specific runoff. In this area the big rift valleys opened to the North have a very significant role. The humid air masses can move from these valleys into the interior of the mountain without orographic rise. In the mountain territories behind the valley entrances at the same altitude the specific runoff is greater than in the Central and Eastern parts of the Alps.

58

3. The third type is characterized by the central part of
the Alps. The absolute values of the specific runoff keep on
decreasing in comparison with the other two types. The vertical gradient of the specific runoff also reaches zero. This
type can be observed in the river-system of the Drava, Mura
rivers. The lower air-masses arrive at these regions after repeated orographic rises, thus their degree of humidity is
very much reduced. At the same altitude higher specific runoff now here can be found but in the region of lower passes
providing connection with the Northern ridges of the Alps.

4. The fourth type comes about on the lees of orographic
systems. This is the case with the regions of the Northern
Carpathians running down towards the Carpathian Basin. The
absolute values of the specific runoff in these territories
are even less, than that of the former type. The vertical
gradient of the specific runoff does not show significant
difference from the one found with the third type. This type
is represented by the river-systems of the Vág, Nyitra, Garam,
Ipoly. The lowest value of the specific runoff among all the
types described hitherto is due to two factors. The lower air-
masses reach the above mentioned river-systems after repeated
orographic rises, thus their degree of humidity is very low.
The geological configurations of the terrain in these river-
systems are rather infavourable, that is why there is a great
loss of evaporation and sweating. Owing to these two factors
the specific runoff is exceedingly little at different altitudes.

5. The fifth type can be observed in the Northern ridges
of the Dinaric mountains. The causes are nearly identical to
that of the fourth type. But because of the determining role
of the geological configurations it seems to be necessary to
describe it as an independent type. Its characteristic feature
is the extremely colourful vertical system of the specific
runoff. This type can be described as a mixture of all the
above mentioned ones. Generally characteristic of this type
is that both the vertical gradient and the absolute values of
the specific runoff are decreasing from the North-West (Száva

river-system) to the South-East (Morava river-system). The
vertical tendencies of the values in the Száva river-system
are the same as the ones observed in the Northern region of
the Alps. However the causes are different. In the Száva
river-system the air-masses that come from above the Adriatic
without precipitation have a significant hydrological role.
The South-Western air-masses, before reaching the Száva river-
head, pass over the 30 - 40 km wide Karst only once. They do
not go dry to a large extent. To the South-East, the air-
masses arriving at the Vrbas, Bosna and Drina river-systems
have to pass over an increasingly wide mountain block, with
the result that their degree of humidity is gradually reducing. This phenomenon explains the gradually decreasing absolute values of the specific runoff in the vertical. The least
one is in the Morava river-system, where the air-masses bring
the smallest degree of humidity, in consequence their hydrological influence is the most insignificant.

Mora N.

(CUBA, Havana)

DECIMAL CLASSIFICATION OF CUBAN RIVERS

This decimal classification has been made for the purpose
of identifying all the rivers of the country, together with
each one of their tributaries, or at least those which have
certain importance within the drainage network.

To identify a river by its name shown in the topographic
chart (scale 1:50 000) brings about many difficulties. The
following are the most common:

Rivers with the same name in different natural regions
and even in the same region; long rivers with no name at all;
different names for the same river along its whole course,
etc. All of these reasons made us think of adopting a decimal
classification for our rivers. And for this purpose, we analyzed the Soviet Method as well as Horton's, Strahler's, Milton-
Oiler's and Storet's Methods. The result of these analyses was

that none of them were useful because they were not adapted
either to our specific conditions or to our purposes.

For the reasons already mentioned, we were compelled
to establish a decimal classification method for our rivers,
which not only fulfilled our purpose but also adapted to our
conditions. We wanted to obtain such a classification that,
with a few numbers and in a simple away, gave us the greatest
amount of information possible about a specific river.

Description of the method developed

In Cuba there are approximately 532 basins, 240 in the
north watershed and 292 in the south one. No matter how much
these figures increase, in our classification, they will never
exceed the number of 500 in either of the watersheds.

We decided to number the basins from west to east and
also by watersheds to avoid the use of large numbers. In the
same way, we numbered the natural regions from west to east
and the north watershed has the number 1 and the south watershed, the number 2. For the natural region of Isle of Pines,
the watershed was "zero" because the drainage network was
mainly radial there. If we take, for example, the river of
the basin marked with an arrow on this map, its code will
be 11.005.

The first digit indicates the natural region and the second, the watershed. Both of them will be separated from the
rest by a decimal point. The three digits following the decimal point (005) identify the basin of the river involved.

In our classification, we took a minimum basic area of
5 km^2 because the majority of the basins of the Sierra Maestra
south watershed, which are the smallest ones of our country,
have, approximately, this area extension. For the purpose of
selecting the tributary minimum lengths to be classified in
basins of different areas, the drainage density of 18 basins
was studied. These basins were selected throughout the country and they varied in relief from flat to mountainous surfaces and in area from 5 km^2 to 9 000 km^2. After this study was
completed, we arrived at the following conclusion:

1) For areas from 5 to 19 km^2, the minimum length for a

tributary to be considered would be 2 km.

2) For areas from 20 to 99 km^2, the minimum length for a tributary would be 3 km.

3) For areas from 100 to 299 km^2, the minimum length for a tributary would be 4 km.

4) For areas from 300 to 1 999 km^2, the minimum length for a tributary would be 6 km.

5) For areas from 2 000 km^2 and over, the tributaries greater than 10 km would be selected.

The classification of these tributaries within the basins was made as follows. All the tributaries on the right side of the main river from its source to its mouth would receive even numbers beginning with the number 2 and all of the left side would receive odd numbers beginning with the number 1. That is to say, that when we are going to classify a tributary first of all, we write the number of the natural region, the number of the watershed and then, but separated by a decimal point, the number of the basin and finally, also separated by another decimal point, the number of the tributary (that has 2 digits). The secondary tributaries will be numbered in the same way; the number of the natural region, that of the watershed, that of the basin, that of the river of which it is a tributary and finally the number of the secondary tributary, separated always from the preceding by a decimal point. In this type of codification, the last group of digits from the basin number tells us the classified river order, for example, 11.005.02.02 (river of second order). This classification has been adapted already in the western region of Cuba. As an example, we can mention the Mantua River basin. All the facts already mentioned serve to indicate roughly the work that we have done in order to establish a classification of the rivers in such a way as to adequately match the present state of development of our country.

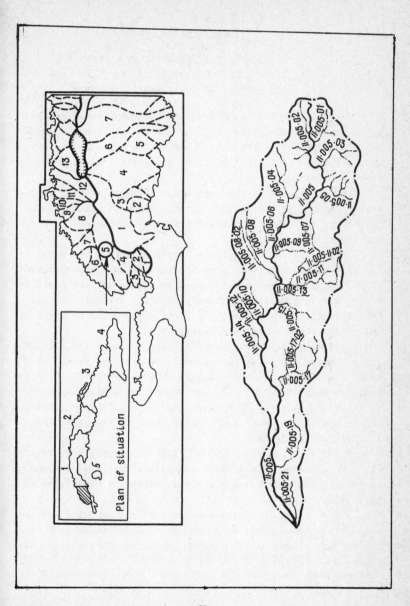

Plan of situation

L.I. S v e r l o v a
(USSR, Khabarovsk)

THE MAIN REGULARITIES OF THE CLIMATE VARIATIONS IN TIME AND SPACE AND THEIR CALCULATION IN THE LONG-RANGE PREDICTION.

The natural conditions are a rather complicated geographical system, consisting of lithosphere, hydrosphere, atmosphere and biosphere. The variations taking place in one of the components indoubtedly involve the variations in the other components.

The atmosphere is the most dynamic component. The atmospheric changes influence the climatic characteristics of the territory. The climatic changes influence structure of flora and fauna, sedimentation.

In process of the geological development of the Earth the fluctuations of such climatic elements as the temperature of the earth and the air, the precipitations and the humidity of the air occured simultaneously with the gradual cooling. Some climatic changes have been registered by the meteorological observations during the modern period.

Studies by E.S. Roobinshtein (1946), A.A. Vittels (1948), L. Polosova (1963) et al. showed, that for the period of time from 1805 to 1960 the climate grew warmer in the Northern Hemisphere in 1922-1945. The temperature rise first occured in the North Atlantic (Greenland) on the territory adjoining the magnetic pole, and then it took place in the other areas of the Northern Hemisphere.

For the period from 1929 to 1965 the highest temperature rise was observed in the North Atlantic (Greenland, Iceland), where the mean annual temperature exceeded the mean value of many years by 2.5°C.

The mean annual temperature on the rest territory of the Northern Hemisphere exceeded the standard value by 0.1-0.6°C. This allows the conclusion, that the fluctuational variations of the climate are first displayed and reach their maximum near the poles and then spread over the hemispheres.

That is why, when considering the problem of the character
of the climatic changes near the poles, we used the long-
range meteorological observations by Gothob station (Greenland) and Orcadas station (Antarctic Continent), represented
in works by E.S. Roobinshtein, L.   Polosova (1966) and
B.Sherdtfegger (1973) and compared the behaviour of the temperature lines in January at Gothob meteostation and the
mean temperature value in July-August at Orcadas meteostation.
Orcadas station is situated in the Southern Hemisphere, latitude 60.7^o, longitude 44.7^o, and Gothob station has nearly
the same geographical parameters, but it is situated in the
Northern Hemisphere.

The temperature fluctuations in the Northern and Southern Hemispheres were developing metachronously from 1905
to 1925. In 1909-1910 in Greenland the air temperatures in
winter were lower than normal and on Antarctic Continent -
higher than normal.

In 1914-1915 the climate warming has noted in Greenland,
and the climate cooling - on Antarctic Continent.
In 1919-1921 Greenland was in the phase of cooling,
and Antarctic Continent - in the phase of warming. During
the following years since 1921 to 1925 the amelioration of
the climate was taking place in Greenland, and it was continuing to grow cold on Antarctic Continent. After the year
of 1925 it began growing cold in Greenland, as for Antarctic
Continent - the cooling there was continuing. One should
note asynchronism in the climate variations.
This phase displacement in time was observed from 1925
to 1942. In 1942 the warming in Arctic Region corresponded
to the warming on Antarctic Continent. After the year of
1942 it was growing cold both in Greenland and in Antarctic
Region. It continued up to 1947. Later the tendency of
growing warmer was observed in the climate of the poles.
According to these data the climate variations near the

poles can be both metachronous, asynchronous and synchronous. The noted fluctuations of the climate occur simultaneously with the more prolonged climatic changes (the phases of warmth in Arctic Region and the phases of cold in Antarctic Region). It is supported by the fact, that the climate in Arctic Region is two times as severe as the climate over the Northern Hemisphere. The mean monthly air temperatures in Arctic Region change during the year from 3.8 to 33.8° C, and in Antarctic Region- from 32.7 to 68.6°C.

The analysis permits the following conclusions:
1. The climate fluctuations first occur and reach maximum in the area of the magnetic poles, and then they spread over the hemisphere while the amplitude attenuates. The intensity of anticyclogenesis over the polar latitudes is constantly changing in time. Cooling is connected with the strengthening of the anticyclogenesis and the increasing number of the arctic invasions into the zone of the temperate latitudes, and warming - with the relaxation of the anticyclogenesis and the decreasing number of the arctic invasions into the zone of the temperature latitudes.

2. Presently the Northern Hemisphere is in the phase of the warm period, and the Southern Hemisphere - in the phase of the cold one, greater cycles in the development of the climate modification. The insignificant variations of the thermal regime near the poles occur simultaneously with these main cycles. They occur metachronously, asynchronously and synchronously.

These regularities must be taken into consideration when analysing the climatic conditions of the past and when forecasting.

The history of the Earth development is connected with its cooling. Thus all the past climatic fluctuations of the different cycles took place on a descending scale (Markov,

1968; Sverlova, 1972). Many thousands years ago the traces of the human activity appeared, they were the burnt and cut woods and the drained swamps. Nowadays such branches of industry as a chemical one and a metallurgical one are rapidly developing . This involves throwing up the waste products into the air, seas and rivers. Thus the chemical composition of the atmosphere and the water (the natural conditions we live in) is greatly changed. This involves "the greenhouse effect" and gradual temperature rise of the lower layers of the air. Thus the human beings have managed to change the thermal evolution of the lower atmospheric layers and direct the climate variations along an ascending scale.

According to Manabe's prognosis (1970) the persistent increase of the concentration of carbon dioxide in the atmosphere till 2000 will cause the temperature rise by $0.5^{o}C$, and till 2070 - by $3.2^{o}C$. The gradual warming of the climate, predicted by Manabe, is periodically interrupted by the natural fluctuations, which now and then blur the role of the antitropogenetic factor in the climate modification and sometimes removes it completely. This is confirmed by the data, obtained from the air temperature observations, conducted by many stations in the Northern and Southern Hemispheres. Thus the natural fluctuations in the climate modification during the present period of development of human society are decisive; they determine the climate variations. As for the human activity, it plays a secondary role. It decreases the natural cooling and increases the effect of the natural warming.

Conclusions

Long-range prediction must be based on the main regularities of the natural fluctuations in the climate modification and the human activity influence on the tropospheric thermal regime.

Figures to the article by Oguntoyinbo J.S. (Volume II, Page 114)
Fig. 2 is not presented here for technical reasons.

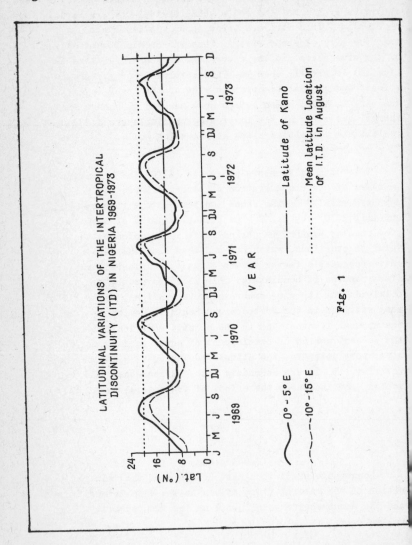

Fig. 1

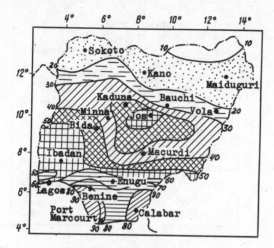

Fig. 3a.   NIGERIA TOTAL RAINFALL 1973
(values in inches)

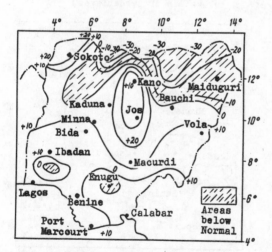

Fig. 3b.   NIGERIA RAINFALL DEPARTURE
from NORMAL 1969
(values in percentages)

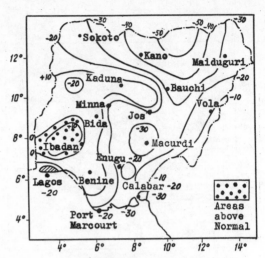

Fig. 3c. NIGERIA RAINFALL DEPARTURE
from NORMAL 1973
(values in percentages)

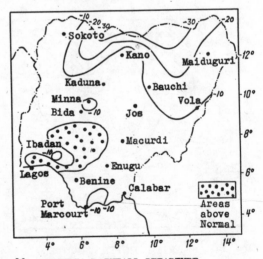

Fig. 3d. NIGERIA RAINFALL DEPARTURE
from NORMAL 1969-73
(values in percentages)

70

III. GEOGRAPHY OF THE OCEAN

Avello O., Ionin A.

(CUBA, Havana; USSR, Moscow)

RELIEF OF THE CUBAN SHELF AND SOME PECULIARITIES OF ITS DEVE-LOPMENT

In the last years much attention has been given to the study of the relief and bottom sediments of the tropical zone shelfs, including the Atlantic ones, primarily because of its relation with the fisheries industry and with the investigation on geological prospecting.

The undertaken investigations showed that the relief of insular and continental shelfs of the tropical zones of the oceans is different from the relief of the submerged borders of the continents and the insular flats which are distributed in higher latitudes. This difference is fundamentaly based on the geological structure which presents complex forms of numerous types. This is related partly with the important role played by the processes of biogenic formation, particularly the reef on the tropical relief and partly with the peculiarities of the constitution of these biogenic forms of relief in the conditions of repeated transgressions and regressions of the World Ocean occurred during the Quaternary and the temperature changes of the water masses.

Cuba is almost completely bordered by an insular shelf which width varies from several hundred meters to several kilometers.

Chains of coral reef barriers and numerous keys, different in their genesis and geologic structure, extend along the border of the Cuban shelf as well as along several parts of the tropical zone of the Western Atlantic.

The undertaken investigation of the Cuban shelf and the analysis of the cartographic materials as well as the available literature on the Western Atlantic zone, show quite clearly that it is because of the peculiarities of the morphogenetic processes (mainly the biogenic ones) of the tropical zones, that these differences significantly disappear and there are similarities in the surface structure and the littoral sedimentogenesis.

For example, shallow waters in the littoral banks are characteristic of Cuba which is situated in the limits of the Antillean Geosinclinal Arch and also of the zones which border the Florida and Yucatan Shelf. In Cuba the depths vary between 6 and 10 m (in some cases up to 20 - 25 m), while in the Yucatan Peninsula these reach 60 - 70 m and in Florida 10 - 15 m.

The shelf border in the tropical Western Atlantic is located at different depths. In Cuba it is at some 20 - 40 m, less frequently at 50 m; in the Mosquito Banks, at 80 - 90 m and in the western part of the Yucatan shelf it reaches a depth of 70 - 100 m.

The parts of the Cuban shelf which present a greater amplitude are: The Gulf of Batabano, which waters occupy the shallow bottoms located between the anticlinals of Pinar del Río province and the Isle of Pinos; the Gulf of Ana María and Guacanayabo, the first located in the sinolinal depression of Ana María while the second coincides with the continuation of the deep depression of the Cauto river; and the Gulf of Guanahacabibes which is at the depression of La Fe.

The narrow parts of the shelf, with an extension of several hundred meters or even kilometers, are located at the flanks of anticlinal structures, as we can observe in the north littoral of Havana-Matanzas.

In the zone of littorals cut by young faults from the Pliocene-Quaternary, there is no evidence of the shelf. For example, along the south coast of Oriente province; in the Peninsula of Guanahacabibes; in Cochinos bay, etc.

72

The biogenic morphosculptural forms of relief are diverse in the limits of the broadest parts of the Cuban shelf and are represented by coral barriers, microatolls, etc.

The surface of the tropical shelf, can be divided into the following types according to the processes of formation of the relief:

1) relict subaereal (karst erosion and eolics);

2) biogenic;

3) recent relicts and hydrogenics (accumulations by waves and abrasion; created by bottom currents of different origin and by suspension currents);

4) potamogenics.

The peculiarities that distinguish the barrier reefs of the Cuba shelf from those of the Indo-Pacific are:

1) the small thickness of the carbonated deposits;

2) their small width;

3) their location, sometimes several kilometers from the shelf border of shallow waters;

4) small amount of coral species.

These findings as well as the composition and structure of the bottom sediments of the Cuban shelf made possible the representation of the development stages of the relief of the shallow littoral zones from the Upper Quaternary.

Gershanovich D.E., Moiseev P.A.
(USSR, Moscow)
PHYSICAL-GEOGRAPHICAL AND ZONAL BASIS OF REGIONAL OPEN SEA FISHERIES

The contemporary world fisheries make use of the main part of the traditional biological resources of the World Ocean, such as large invertebrates, fish and aquatic mammals. In 1969 - 1975 the total catch of marine organisms reached 60 mln tons. The assessments made by the most prominent experts of the field indicate that a further considerable increase of the catch in the ocean would hardly be possible.

The present day level could only be exceeded by not more than 30 - 50%. Therefore, sea fisheries conducted at this stage without any significant introduction of mariculture and various forms of biological melioration, transplantation, etc. are virtually entirely based on the natural biological productivity of the World Ocean, and are determined by its size and distribution. The possible addition exemption of biological material at the expense of Antarctic krill which in theory can be large will only enhance the above mentioned prevailing situation.

The results of research have shown that the biological productivity of the World Ocean is most closely connected with the natural zones of the ocean. Ties with both latitudinal physical-geographical zones and vertical depth zones can be tracked down. Besides, of particular importance, is also the so-called circumcontinental division by zones bearing on the extent of remoteness of a certain oceanic region from continental coasts or island arcs, and on the accompanying changes in the regime and structure of waters, impact of water flow, and character of life. Similar ties are observed in the distribution of nearly all abiotic factors of the marine environment which determine the new formation of organic substances, in the initial links of the trophic chain of sea organisms, as well as in all of its successive links, if we are to consider the whole problem of biological resources in the ocean.

The differences in the new formation of organic substances in the World Ocean are sufficiently great. The primary production can vary from 0.05 to 3.0 grams C/m^2 a day, i.e. by 60 times. We can distinguish between the oceanic areas of high productivity, and a sort of oceanic deserts. Their location in the ocean is much less determined by the latitudinal physical-geographical zones than on land. As a rule, regions adjacent to shores within each latitudinal zone are more productive than the regions more remote from the continents. This leads to the emergence of continent adjacent zones of higher productivity in each ocean and deep seas. In many cases the width of these zones coincides with the width of the shelf and

the neighbouring part of the continental slope. In the open
sea the regions of high productivity always center about the
areas of frontal and upwelling zones, and where the water
strata are less stable. Sharply increased productivity is
characteristic of the areas of intensive coastal upwelling,
such as the Peruvian and Californian ones in the Pacific,
Bengali and Canary ones in the Atlantic, etc., and is primary
a feature of narrow shelf zones. Because of the most intensive
dependence between groups of organisms in the biocoenosis
these peculiarities in the distribution of the primary production also have an impact on the secondary production. Areas
of higher secondary production have limited square comprising
about 10% of the water area of the World Ocean. Coastal upwelling zones make up only about 0.1% of that space. However, it
is exactly these 10% of the oceanic water area where approximately 96% of the total catch in the World Ocean are harvested
from (up to 90% on the shelf).

The north-western, north-eastern and south-eastern areas
located on the sides of the Atlantic and Pacific oceans are of
greatest importance in world fisheries. There are the main
areas of anchovy, cod, herring, mackerel, horse-mackerel, flat
fish and other fisheries. In 1971 the Pacific yield constitu-
ted 55% of the world catch, Atlantic - 39%, Indian Ocean - 6%.
The harvest in the Pacific therefore exceeds that of the Atlantic and Indian together. The share of pelagic fish is over
63%, bottom and near bottom fish making 23% of the total catch.
The remainder are the fish whose proportion to these two main
groups is not very clear.

It is very indicative that all the northern zones of the
World Ocean, where many continental shelves are concentrated,
yield 54% of the ocean catch, whereas the southern zones provide only 25% (the tropical and Equator zones - 21%). Hence,
if one considers the overall regional pattern of the world
fisheries and its correspondence to the natural geographical
zones, he might conclude that the latitudinal physical-geographical zone pattern in the first turn determines the composition of the biological complex, thereby the presence of com-

mercial organisms. The formation of more productive areas of the ocean as a whole is naturally connected with the latitudinal zone pattern. It is responsible to a great extent for the variety'of life forms in the subpolar, temperate and the subtropical waters, for the formation of the trade circulation and, consequently, the appearance of powerful current systems along continents and across the ocean in the equatorial, tropical and partially temperate latitudes many of which are also distinguished for their raised biological productivity. We must accentuate the spacious zone of transference of water along the Antarctic continent having rich content of biogenic elements, phyto and zooplankton.

At the same time these zones of great importance have less impact on the regional pattern of fisheries than the circumcontinental ones. Apparently the latter are greatly important not only because the continent adjacent regions of the ocean are sufficient in minerals and organic substances, but also because of the wide variety of habitats of organisms, which allows for the choice of the most favourable conditions at any stage of their development. We cannot rule out the possible effect of the factors involved in the development and settlement of the organisms.

This allows up to specify the following fishery zones of particular importance within the world fishery area pattern:
1) Coastal fishery.

2) Shelf and close neretic fishery.
3) Remote neretic fisheries.
4) Open sea fisheries (including epipelagic fisheries, and the elevated areas of the ocean).

Each of these zones can be subdivided into subzones or regions affected by the factors which raise biological productivity. For example, estuaries, bays, and lagoons become particularly important for the inshore fisheries. They also have some effect on the shelf fisheries. In that case special role is played by the areas of upwelling, whirls, places of higher convective and turbulent mixing of waters, etc. In the

76

oase of remote neretic fisheries one has to take into account
the position of currents and their separate streams, meanders,
local whirls, upwelling points, frontal zones. Together with
upwelling in oceanic whirls and vergent zones, the same factors determine the regional pattern of oceanic fisheriés. We
must point out that the role of latitudinal zones is especially clear in the open sea fisheries.

The connection between many definite fishing grounds and
zonal anomalies, local abnormal water regimes and the accompanying fluctuations in the abundance and distribution of organisms is observed everywhere. Their permanent monitoring,
tracking down period regularities and, in many cases, their
forecast are of primary theoretical and practical importance.

Holmgren, B.E.
(SWEDEN, Uppsala)
TURBIDITY OF THE ARCTIC ATMOSPHERE IN RELATION TO OPEN LEADS
IN THE PACK ICE

Introduction. Several investigations of the atmospheric
turbidity during cloudfree conditions in the Arctic show peak
values during the spring period. Open and freezing leads and
polynyas in the polar pack ice release as yet unknown quantities of heat and moisture into the atmosphere during the cold
period of the year. In a subproject of the AIDJEX[x] Lead Experiment during March 1974, the possible role of open leads in relation to water vapour entrainment into the lower troposphere,
ice crystal formation and subsequent increase in the turbidity
was studied in the Barrow area in northern Alaska. The temperature static stability close to open leads was monitored by
a vertically pointing sodar[xx]. Simultaneously, the total atmospheric turbidity was measured by a Linke-Feussner actino-
meter, and the vertical distribution of the Mie scattering

---

[x] AIDJEX - Arctic Ice Dynamics Joint Experiment

[xx] sodar - sound detecting and ranging

coefficient by an airborne photometer. Ice crystal densities were measured by various sampling devices. Details as to the instruments and results have been presented elsewhere (Shaw, 1975; Ohtake and Holmgren, 1974; Holmgren, Shaw and Weller, 1974; Holmgren and Spears, 1974).

Boundary layer structure and vertical humidity profiles. The turbulent structure of the atmospheric boundary layer was continuously monitored by the sodar during the entire month of March. The sodar antenna was a vertically pointing parabolic disk used both for transmitting and receiving. The back-scattering signal is then related to the small-scale temperature fluctuations only. The lowest probing range of the sodar was 25 m and the highest range 650 m. A well-defined statically stable mixing layer was found about 60% of the recording period. During the rest of the period the top of the mixing layer was probably too low to be recorded, i.e. below the lowest probing range. In the air of high static stability above the mixing layers, the acoustic soundings indicated wind shearings and turbulence confined to thin sheets between layers of little or no turbulence. On two occasions, air was advected across newly opened leads off the coast and then passed the sodar site, which was situated 2 km inland. Both times the depth of the mixing layer expanded from 100 m to a maximum of 250 - 300 m. After the closing of the leads, the mixing layer depths decreased to the original values of 100 m, indicating that the convection above the leads caused a considerable expansion of the mixing layers into which heat and moisture may be effectively transported.

An analysis of the routine radiosondings made twice daily at Barrow showed that practically all of the humidity profiles during March 1974 had markedly lower mixing ratios at the surface than at an elevation of a few hundred meters. The average mixing ratio at the surface was 0.33 g/kg, and the average maximum mixing ratio was 0.86 g/kg at an average elevation of 860 m. Sample statistics of the humidity profiles in March showed maximum mixing ratios at levels below 600 meters in 50% of all cases. The cold snow surface obviously serves almost

78

continuously as a sink for atmospheric water vapour during
this time of the year. Except for open leads there are no major atmospheric moisture sources within the Arctic Basin. Referring to the results of the sodar measurements, it does not
seem out of the question that large polynyas may contribute
to the maximum mixing values found at low elevations.

Atmospheric turbidity and ice crystals. The Ångström turbidity coefficient ß in cloudfree conditions appeared to be
0.065 calculated as the average for 22 days in March. This result may be compared with an average β-value of 0.085 for
10 days in the Beafort Sea during April, 1972 (Weller et.al.,
1974). The incidence of visible ice crystals in the surface
layer and peak turbidity values were in both series strongly
correlated. On some days the ice crystal formation related to
nearby open leads was obviously connected with the peak turbidity values. During numerous flights over the pack ice
north and northwest of Barrow, pronounced haze layers in the
lower troposphere occasionally reduced the horizontal visibi-
lities to the order of a few kilometers, and the haziness
could in fact resemble something one might encounter over big
metropolitan areas in the south. Haze above the lowest kilometer or so often had a brownish-yellow tinge. Vertical profiles of the aerosol volume extinction coefficient made by an
airborne photometer in cloudfree conditions showed pronounced
peaks corresponding to these visually observed haze layers
with maximum values of the volume extinction coefficient surpassing that of the hazefree air by a factor of 15. A comparison between the visual observations made during the flights
and the radiosond humidity profiles showed that the haze above
the mixing layer often corresponded to relative humidities
greater than 70 - 80%, indicating that the haze might be caused by water vapour condensation on hydroscopic nuclei. On
the other hand, there was also haze encountered where the corresponding humidity was as low as about 50%.

Discussion. The observations showed that the concentrations and the types of aerosols within the mixing layer were
generally markedly different from the aerosols above the mix-

ing layer and that the vertical aerosol distribution was critically dependent on the vertical temperature, humidity and wind variations. If water vapour from the wide leads can escape via convection into the stable layers above the unexpanded mixing layer, the residence time for water vapour in the atmosphere should increase drastically as compared to moisture transport limited to lower elevations only. In the former case one might expect that the effects of the increased humidity on the haziness and ice crystal formation may be noticeable far away from the leads, i.e. not only on a local scale.

In conclusion, there is no doubt that the peak turbidity values at Barrow were associated with ice crystal aerosols which in some cases could be directly related to nearby open leads. However, turbidity values of about 0.050 and higher were measured on several days when neither the radiosond humidity profiles nor the visual observations indicated the presence of ice crystal aerosols. Evidently, the aerosol densities above the pack ice, causing significant turbidity variations, cannot be ascribed to any one single factor.

On the average, the observed turbidity values are too high to be ignored in radiative energy budget calculations of the Arctic atmosphere. Further studies involving conventional pyrheliometry, sample collection and analysis of the chemical composition of the aerosols, determinations of vertical aerosols profiles, analysis of the aerosol size distribution, satellite imagery to record the frequency of occurrence and size of open leads, are needed before a proper assessment can be made of the role of aerosols in the Arctic energy budget.

Monin A.S.
(USSR, Moscow)

ABOUT THE NEW MULTIVOLUME RESEARCH IN OCEANOLOGY DONE IN THE SHIRSHOV INSTITUTE OF OCEANOLOGY OF THE USSR ACADEMY OF SCIENCE

The "Oceanology" series offered for publication has no equal in the World literature for the coverage of modern oceanographic problems and systematic presentation of the material. The monographs comprising the series sum up to many-year achievements of the World oceanography as an integrated science and give an adequate basis for the development of future scientific programmes.

The "Oceanology" series consists of 10 monographs of 35 sheets each.

Volume 1, "Hydrodynamics of the ocean", presents the foundations of sea water thermodynamics. Gravity waves - wind-driven, internal, tsunami and tidal are dealt with in detail. A separate chapter is devoted to synoptic gyres in the ocean. The monograph is concluded by a consideration of the general circulation problems, including numerical modelling of the ocean currents.

Volume 2, "Oceanic physics", deals with oceanic turbulence. Peculiar microstructure of the ocean is considered in detail and turbulent diffusion of admixtures is discussed. The interaction between the ocean and the atmosphere is dealt with as an all-important geophysical problem. Oceanic acoustics, optics and nuclear hydrophysics are expounded in conclusion.

Volume 3, "Marine chemistry", presents a detailed consideration of the chemical composition of ocean water. Chemical pollution and chemical resources of the ocean are discussed. Variability of the chemical regime of sea water, rates and scales of biochemical, physico-chemical and chemical processes in the ocean depths and at the ocean-atmosphere and the ocean-land boundaries are investigated. The concluding section gives an idea of the annual cycle and chemical balance of the World Ocean.

Volume 4, "Chemistry of sediments", is devoted to the formation of chemical composition of sediments. Consideration is being given to changes of physico-chemical conditions in diagenesis compared to sedimentation conditions.

Volume 5, "Ocean floor geophysics", gives a detailed description of the up-to-date methods employed in marine geophysical research. Physical fields and deep lithospheric structure of the earth's oceanic regions are investigated.

Volume 6, "Geodynamics of the ocean", contains modern concepts on the dynamics of tectonic transformation of the earth's oceanic regions and the oceans' fate in the global development of our planet.

Volume 7, "Marine geology" gives basic information on the geological structure of the World's oceans, on stratigraphy, lithology, mineralogy, geochemistry and sediment facies from the recent ones back to Upper Jurassic. Geological laws governing the distribution of hard mineral resources (polymetallic iron-manganese rocks, metalliferrous sediments, phosphorites, etc.) over the ocean floor are considered.

Volume 8, "Geological history and tectonics of the oceans", contains an analysis of the available data on the history of the geological development of the oceans in Mesozoic and Cenozoic time, including the deep-sea drilling results; sedimentary and magmatic formations are described. Generalizations are presented in the paleogeography and tectonics of the earth's oceanic regions.

Volume 9, "Biological structure of the ocean", presents a description of the major environmental factors and the problems of adaptation of marine organisms. Peculiarities of the biological structure and the biophysical fields of the ocean are discussed.

Volume 10, "Biological productivity of the ocean", considers the ecology of marine communities (ecosystems). The use by man of the biological resources of the ocean is dealt with. Anthropogenic pollution effects upon marine communities are discussed.

The "Oceanology" series is of an exceptional interest for all oceanographers: physicists, chemists, geophysicists, geologists and biologists. Separate volumes of the series will undoubtedly be useful to wider circles of physicists, chemists geologists and biologists who can find in them new problems, new ideas and new solutions. The publication will evidently be helpful for specialists of hydrometeorological service, commercial fishery, navy, merchant and passenger fleet, practical geologists, engineers and other specialists whose work is con-

nected with the seas and oceans. Finally, the monographs of the
series will be valuable textbooks for students and post-gradua-
tes of universities, hydrometeorological, physico-technical,
fishery and geological institutes.

Nazarov V.S.
(USSR, Moscow)
THE ICE COVER OF THE OCEAN - THE INDICATOR OF THE EARTH CLIMATE

The World Ocean regulates the heat reserve and its distribution on the Earth. The seasonal and climatic variations are
determined by the quantity of the sun heat, coming to the
Earth, and its redistribution proceeded by the sea and atmosphere currents.

Depending on the rise and fall of temperature of the upper layer of the ocean water the ice cover of the World ocean
spreads more or less, respectively, from the polar regions to
the equator, while not only the area of the ocean surface covered by ice decreases or increases, but usually the thickness
of ice and therefore ice mass in the sea change the same way.
So, ice spreading in the ocean or variation of its mass characterize quantitatively the heat state of the Earth surface.
The ice cover of the ocean sums up the influence of the all
factors generating the climate and gives the numerical result
in form of the ocean and sea areas covered by ice.

There is a close relation between the variations of the
ice area in the Barentz Sea and the water temperature on the
Kola Peninsula meridian and the correlation coefficient between these variables equals to 0.76 and between their extre-
mal values - 0.98. The ice area in the Kara Sea in the first
half of September is rather indicative value of the climatic
variations taking place.

The seasonal variations of the mean latitude of the ice
spreading boundary are 5° in the northern hemisphere and 10°
in the southern hemisphere; interannual variations of the ice
spreading boundary are 2° and $3-4^o$, respectively.

As the climate variations cover the large areas of the
earth surface, have the global character and occur simultane-

ously in the time scale of climatic variations, that is shown in works of V. Vize and V. Nazarov, one can trace the long-term course of variation of ice area in the ocean according to the data of one sea.

In 1947 V. Nazarov investigated the variations of ice area in the Kara Sea in the end of summer from 1750 to 1947. For forecasting of possible climatic variations to 2000 the reliable data of water temperature variations on the meridian of the Kola Peninsula and of ice area in the Kara Sea from 1925 to 1975 were used.

Extrapolating the course of the integral curve of ice area of the Kara Sea for the one forth of the period to 1990 - 1995 one can expect that to 1990, as a rule, the ice area in the Kara Sea will be more than its mean value, in 1990 - 1995 the ice area will be mean and later on - less than mean value. Therefore, in the World Ocean in the northern and the southern polar areas the ice distribution towards equator will be, as a rule, more than mean to 1990, i.e. the fall of the temperature on our planet will take place. As a result of forthcoming climatic variations some rising of the level of the Caspian Sea and, as the melting of glaciers will be dece-lerated, the descend of the Aral Sea level can be expected. In the mean latitudes the continental climate will remain, the summer will be more hot, and winter - cold, and the trans-fer of the air masses with sharp changes of the temperature and wind regime will be mainly zonal in summer and meridional - in winter. Likely, the increased moisture of the Sevan Lake region and the part of India, and increase of the Earth glaci-ers will remain.

Rassokho A.I.
(USSR, Leningrad)

CONTRIBUTION OF THE SOVIET NAVY TO THE DEVELOPMENT OF THE
WORLD GEOGRAPHICAL SCIENCE

Participation of the Soviet Navy in the development of the geographical science has old well established traditions. Already in the XVIII century all commanders of the Russian

naval ships bound for long-term sailings had a standing order
to collect while under way data relevant to physical geography
of seas and oceans and include the data in their reports.

In 1803 - 1806 the first Russian round-the-world expedition was undertaken under the command of I.F. Krusenstern. The
scientific results of the expedition summarized in 1824 in the
"Atlas of the South Sea" acquired world-wide fame. In the years
to follow round-the-world sailings of Russian navy men became
a customary occurrence and many of them led to geographic dis-
coveries.

Of great importance were sailings under the command of
F.F. Bellingshausen and M.P. Lasarev, O.E. Kotzebü, F.P. Lütke.
World-wide recognition has been won by the works of Admiral
S.O. Makarov, whose book "The 'Vityaz and the Pacific" is in-
cluded in the capital fund of the world geographical litera-
ture.

In the end of the XVIII and the beginning of the XIX centuries in the Russian fleet new detachments were established
responsible for marine survey and study of hydrometeorological
regime of the seas bordering on Russia. These detachments were
further developed after the victory of the Great October Revolution, when hydrographic researches entered the phase of all-
state planning.

By the beginning of the Great Patriotic War of 1941 - 1945
the USSR Navy has mainly completed research in the coastal waters of our country and was ready to start ocean exploration,
but the war intervened.

During the Great Patriotic War hydrographers, oceanolog-
ists and meteorologists of the Navy along with all the Soviet
people struggled valiantly to gain victory over the dark forces of fascism.

After the end of the Great Patriotic War the Navy continued research in seas and oceans. Since 1957 research vessels
of the USSR Navy have taken an active part in fulfilment of
international oceanographic programmes.

Together with vessels of the USSR Academy of Sciences
they took part in researches according to the programme of the

International Geophysical Year and the Year of International Geophysical Cooperation. In 1959 - 1965 Soviet Navy took part in the International Indian ocean expedition. Since the late 1960's there have been practically no large international oceanographic expedition without participation of ships flying the Soviet hydrographic flag.

The USSR Academy of Sciences and Navy regularly exchange results of their researches with other countries studying the World Ocean. Resulting from this cooperation a great volume of oceanographic observations has been accumulated in the USSR. In the late 1960's and early 1970's this volume for the first time was processed by uniform methods which made it possible to reduce to a uniform system and compare within it oceanographic, meteorological and aerological observations made in the World Ocean in all the history of its studying.

As a result of this work the USSR Navy in cooperation with the Academy of Sciences and other leading research institutions of the USSR proceeded to compiling a fundamental scientific geographical work, the World Ocean Atlas.

The first volume of the World Ocean Atlas, concerned with the Pacific Ocean, has already been published. The further volumes of the World Ocean Atlas will be issued over the next few years.

In the Atlas practically all Soviet and the majority of foreign oceanographical and meteorological observations are used, collected in the World Ocean in all the history of its exploration (around 7 million observations).

Analysis of these data as a whole, done for the first time in the world, allowed to avoid subjective assessments and extrapolations characteristic of many existing atlases.

Whereas other atlases describe only the surface layer of the ocean and the immediately adjacent atmosphere, the new atlas contains characteristics of the water medium up to a depth of 5,000 metres and atmosphere up to a height of 16 - 18 kilometres.

Much information is published in the new atlas for the first time also because at the time of issue of the existing atlases it was not yet available.

Such information includes characteristics of the upper layers of atmosphere, deep-water currents, hydrochemical composition of water, bottom sediments, tides in the open ocean, etc.

The first volume of the World Ocean Atlas contains 287 charts in colour, presenting comprehensive up-to-date information about the bottom of the Pacific Ocean, climate, hydrological, hydrochemical and biogeographical conditions..

Information contained in the Atlas covers altogether nearly 900 scientific subjects.

The section "Ocean Bed" gives a thorough investigation of geological structure of the bottom and shores of the Pacific. Besides bathymetric and geological charts this section includes detailed information about earthquakes and volcanoes, bottom sediments and types of coasts. Heat flow through the ocean bed is investigated, which is one of the most important elements of the Earth's energetics characteristic.

The section "Climate" deals with heat and water regime of the Pacific, winds, atmospheric precipitates, visibility, atmospheric circulation and other climatic elements.

"Hydrology" is the largest section in the Atlas. It consists of charts of temperature, salinity, water density and sound velocity. Much attention is given to surface and deep-water currents, water level fluctuations and tides.

The section "Hydrochemistry" contains the most up-to-date scientific data. For all the Pacific charts are made showing distribution of dissolved oxygen, biogenous elements, hydrogenous ion concentration and alkalinity. They cover not only the surface, but also deep layers of ocean waters.

For specialists and all those interested in fauna much information is given in the section "Biogeography". Distribution of plankton, seaweeds, fish, pinnipeds, whales, seabirds and paths of their migration - such is a far from complete list of subjects of this section.

The first volume of the World Ocean Atlas ends with the section "Reference and Navigation-geographical Charts", comprising detailed data on geomagnetism, astronomic phenomena, sea and air communications, population, medical-geographical conditions, etc.

The issuing of the Soviet World Ocean Atlas is an important event in the world geographical science, concluding the first phase of the international cooperation in the study and exploitation of the World Ocean. The Atlas testifies convincingly to the great scientific contribution of the USSR to the knowledge of our planet's nature. The Soviet navy men consider it a great honour to take part in a work directed eventually to the benefit of mankind.

Yamaoka, Masaki
(JAPAN, Kofu)

PROCESS FROM LAND TO SEA ALONG THE JAPANESE COASTS

Along the coasts of the Japanese Islands, we find many settlements with various features of coastal life, which are exhibited at the frontiers of the two different worlds, land and sea, meet.

Role of Land in Coastal Life - The land provides inhabitants with the fields for food production, home-lots for dwelling, forests for fuel and building materials, and graveyard. In Otsurutsu, a small hamlet of 5 houses, they have kept their life for over 700 years, restricting the number of families, sharing equally their lands, aiding each other in co-operative work; and they have never been engaged in fisheries, turning their back to the sea. Many such isolated pure farming settlements with land enough to support the inhabitants are the proto-type of the process from land to sea.

Role of the Sea - The sea has different features as a working place from that of land. 1. It is an unstable and often dangerous place, terra incognita. An unexpected change of catch resulted in a sudden flourish or decadance of settlements in Japan as well as in other countries. Ecological features of important pelagic fish are not known yet. 2. Apart from the international problem, the sea is open to every person in his own national area, at least. Although in Japan, there is the fishing right of water-front belonging to a settlement, the products have been thought as a common property of all inhabitants and they have observed the rule to keep the area productive.

Meanings of Fishing Techniques – To operate fisheries on the unstable and dangerous sea, superior techniques are the only weapon to work in such open area. Hence, the attitude of secrecy to keep the techniques from others. Because, if others should know the very secret fishing places, it means that one might lose his own property to feed his family. There is a rule in some villages to expatriate a person who has let out a secret to a person of another village. Such an attitude remains not only in fisheries of earlier stages but even in modern fisheries. It is important that the attitude and adhesion to the traditional techniques often lead to a stagnancy, working as an inertia, and it is a factor to bring about an unstable economic situation, depending on a gamble-like industry.

Normal Process from Land to Sea – In spatially limited area of Japanese coasts, they have sought the way to reduce the increasing population pressure on land. Fisheries have played a role as an outlet for the surplus population. 1. Map of Fishermen Farming Ratio shows how many fishermen hold and cultivate the fields in each village. Reading the map, we find that points of high ratio are prevalent along the whole coast; particularly isolated islands, which might be considered to be fishery dominant, show generally very high ratio. Except points shown in black, all the villages have the relation with the land, the ratio differing due to each situation. 2. Map of Advancement of Fishing Techniques – After analysing the elements in technological advancement, we use here the biggest boat in tonnage as the criterion for advancement. As the table below shows, with the increase in boat tonnage, the advancement to the sea occurs, accompanied by other variables. Looking at both maps together, we find the following: 1. Both coincide, the farming ratio high and advancement slow, where the points stand far from the consuming centres and need considerable self-supporting life. Speaking generally, the technological advancement goes on with the lessening tendency of land dependency. However, this does not run continuously. In some places, while the farming ratio lowers, the advancement stops. There is a certain stage in advancement which stays in a stagnant situation, sticking to the traditional techniques under poor conditions, hoping for a big catch tomorrow morning. Such a stage in Japan is "koduri" fisheries, small-scale and individual ones. And advancement to the next higher stage is accelerated by other elements-complex. Here, we must add another special factor: taste or liking of consumers. For example, in the Inland Sea area, with the oldest history of fisheries, they stay in the most underdeveloped stage in Japan, surrounded by big consuming population, who favor the delicacy of the marine products with exquisite cooking. Such situation of this area quite resembles that of the Mediterranean area with its role in European civilization. Thus, we can see a genetic process from a pure farming settlement to a modern big fishing port. As the Table shows, each settlement is inhabited by farmers, fishing-farmers, farming-fishermen, or by fishermen. Moreover, in a port, many kinds of fisheries are carried on by various types of vessels. In Misaki, one of the biggest fishing ports in Japan, you see small rowing boats to do gathering of "miduki", a number of small boats under 3 tons to angle squids, 10-20 tons mackerel angling boats and modern big tuna boats over 500 tons to go out to the Atlantic. Such coexistence is seen also at many fishing ports in other countries, such as Grimsby, Fleetwood, Concarneau, Douarnenez, New Bedford, etc.

Reversal Process - There are settlements where fishermen have almost no land to cultivate. These settlements are formed by immigrant fishermen, seeking good demands and superior techniques to carry on commercial fisheries. Castle-town ports were made to serve the new-born big population in the feudal age. To Mimase, a fishing port made for Kochi, many fishermen were summoned by the lord's order, and also some joined to do fisheries in privileged area. Almost all the castle-towns

had such fishing ports near them. In Hokkaido and northern
Kyushu, many new ports were born for the northern ocean fisheries and the East China Sea trawling since Meiji era. Such
fishermen migration occurs in other parts and in other countries too. These emigrant fishermen have their mother settlement, which had experienced the normal process before their
departure to make the second process. Such movements always
expand the techniques to the new places.

    <u>Abnormal Process</u> - The third one has quite a different
origin from the above two. Very few of the black points in
Fig. 1, belong to the group showing the abnormal movement.
Different from those of second process, which is an extension
from the normal one, these people had no relation with land
from the very beginning. In the western part of the Inland
Sea and northern Kyushu, there are boat-people called "ebune"
group, engaged mainly in fishing. They had neither land nor
fishing right and the marriage with them has been a taboo for
neighboring villagers. They have been despised by others and
regarded as in an outcast-like situation. Along other coasts
in Japan, such segregated group, but not boat-people, was neither
admitted to own land nor allowed to do fishing until the Meiji
emancipation. However, they are true Japanese, not other race.
Boat-people in South China and the sea nomads in SE Asia, different from the ebune group, belong to another race than the main
race, seem to be a product of long unstable situation. Different from the central or North China, as the coastal area of
South China has a very unfavourable land condition, they seem
to be obliged to separate from their mother land early. As Kor-
rigan says, the boat-people of Yangtze Kiang have a close relation with paddy cultivation. As there are many things to make
clear, we cannot define them hastily. The despising, not only
to boat-people, seems to be somewhat common attitude to fishermen by other groups. The "fish-eaters"in Europe, and low caste
fishermen in Ceylon might arise from their living conditions, depending on unstable working place. Passive or active,
the life of coastal settlements struggling with terra incognita, must be examined from the land-and-sea relation.

Fig. 1. Fishermen Farming Ratio, 1963

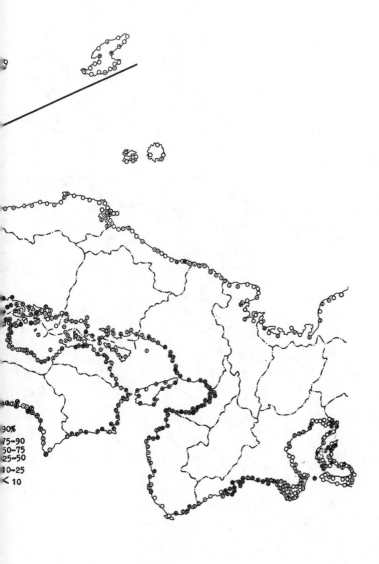

90%
75–90
50–75
25–50
10–25
< 10

93

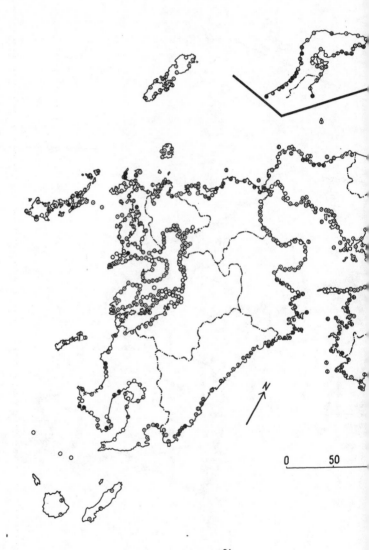

94

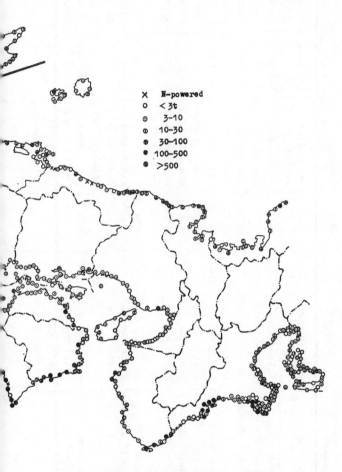

Fig.2. Advancement of Fishing Techniques, 1963

95

Table

Combination of Fisheries Elements

Stage in technological advancement	Fishing method	Craft	Catch	Operation days	Relation with land
I Gathering	Pêche à pied	No	Seaweeds, shellfish	1-3 hrs.	Farming mainly
II Diving	Skin-diving	No	Seaweeds, shellfish	3-5 hrs.	Farming mainly
III "Miduki" fishery	Gathering from overboard	Rowing boat	Seaweeds, shellfish	3-5 hrs.	Farming-fishing
IV Water-front fishery	Beach seine, gill- & lift-nets	Rowing boat	Sardine, spiny lobster	3-5 hrs.	Fishing-farming
Va "Koduri" fishery	Miscellaneous lines & nets	Rowing boat & powered under 3t	Miscellaneous demersal & pelagic fish	1 day	(Woman-farming)
Vb "Koduri" fishery	Miscellaneous lines & nets	3-10 t	Miscellaneous demersal & pelagic fish	1 day	(Woman-farming)
VI Off-shore fishery	Skip-jack p.l., squid angl., small trawl, gill- & purse-nets	10-30 t	Skip-jack, squid, mackerel, trout, frigate mack., etc.	1-3 days	(Woman-farming)

Stage in technological advancement	Fishing method	Craft	Catch	Operation days	Relation with land
VII Off-shore to high-sea fisheries	Skip-jack p.l., squid angl., tuna l.l., med. trawl, salmon drifter, saury s.h.dip net, purse seine	30-100 t	Skip-jack, tuna, squid, saury, pollack, mackerel, salmon, trout, croaker	1 day - 1 month	
VIII High-sea fishery	Tuna long line, purse seine, trawl, drifter	100-500t	Tuna, croaker, mackerel, pollack, salmon, trout	1-6 months	
IX High-sea fishery	Tuna l.l., trawl, whaling	over 500 t	Tuna, pollack, salmon, trout, whale	over 3 months	

AIVAZIAN A.D., PERELMAN A.I.
(USSR, Moscow)

HUMIDLY CATIONIC[x] AND ARIDLY ANIONIC[xx] SPECIES OF PLANTS AND
THEIR CONNECTION WITH GEOCHEMICAL FEATURES OF LANDSCAPES

The distribution of microelements in plants of altitudinal belts of the South-West Altai was studied in zones of ore
deposits (such as polymetallic, copper and chromite) as well
as in ore-free (barren) zones.

The analysis of nearly 10,000 test samples revealed the
peculiarities of distribution of microelements in plants.

Growing in the same altitudinal belt, systematically related plant species are characterized by similar microelement
concentration. Sambucus sibirica, Lonicera altaica and Viburnum opulus of the Caprifoliaceae family are characterized by
low microelement concentration, while Spiraea media and Rosa
acicularis of the Rosaceae family show high level concentration of microelements.

Concentration of microelements in various plant species
is much influenced (besides systematic alliance) by similarity
of history and ecological conditions of these species.

Catiogenic microelements (such as Pb, Cu, Zn) have increased migration ability in landscapes of humid climate with
acid soils (and often waters), while aniogenic microelements
(Mo, V and others) are of low migration ability in these conditions. Thus, it can be assumed that the species formed in

---

[x] The term first introduced by the authors (Russ. - гумидокатные)
[xx] The term first introduced by the authors (Russ. - ариданитные)

humid landscapes are enriched by catiogenic microelements and they faintly absorb aniogenic microelements. This ability of theirs, secured by heredity, most probably appears also in arid landscapes. Such plant species we propose to name _humidly cationic_ (гумидокатные). These include many trees and shrubs, grasses and green mosses of the mountain - taiga, mountain - meadow - forest, mountain and piedmont forest - meadow - steppe altitudinal belts of the South-West Altai.

The ability of a species to accumulate catiogenic microelements, expressed in summarized abundance ratios, we propose to name biogeochemical activity (BCh-catiogenic).

In arid landscapes soils and waters are characterized by neutral or alkaline reaction in which conditions Cu, Pb and Zn show slow migration, while Mo, V and others migrate intensively.

In this connection we assume that plant species formed in arid landscapes are enriched with aniogenic microelements. For these plants we suggest the term _aridly anionic_ (аридаНИТНЫЕ). These are semi-shrubs of the flatland and piedmont desert-steppe zones, such as Atriplex cana, Kochia prostrata.[x]

The ability of a species to accumulate aniogenic microelements, expressed in summarized abundance ratios, we also propose to name biogeochemical activity (BCh-aniogenic).

The conducted researches confirmed that the majority of catiogenic microelements in plants of barren zones is specific for meadow-forest and forest-meadow-steppe belts, while aniogenic microelements are specific for arid-steppe and desert-steppe belts. Thus, the concentration of microelements in the same species of a barren zone varies depending on the altitudinal belt.

a) Highest concentration of catiogenic microelements in the same humidly cationic (гумидокатный) species of barren zones is specific for mountain meadow-forest and mountain and

---

Here we also distinguish a transitional group of plants. Some species of this group are close to humidly cationic, while others - to aridly anionic.

piedmont forest—meadow—steppe belts, i.e.,in these landscapes
the biogeochemical activity (BCh—catiogenic) is of the highest
level.

b) For aridly anionic (ариданитные) species,biogeochemical
activity (BCh—aniogenic) is specific for forest—meadow—steppe
belt; it is somewhat increasing in desert—steppe belt, where
some catiogenic microelements (such as Cu, Zn and Ag) become
more mobile in alkaline medium.

c) Concentration of aniogenic microelements in humidly
cationic (гумидокатные) plant species of ore—free deposits in
mountain—taiga and mountain meadow—forest belts is comparatively low, while in mountain and piedmont forest—meadow—steppe
belt it slightly rises.

d) In aridly anionic plants, aniogenic microelements abun-
dantly concentrate in all steppe belts.

In ore—manifestation areas the accumulation of microelements depends on such factors as the type of ore—deposit, anta-
gonism of microelements, and their forms. For example, in chromite ore manifestation of a desert—steppe belt Artemisia frigida and Stipa capillata highly concentrate Cr, Ni and Mo.
Spiraea hypericifolia accumulates these elements in smaller
quantities. In mercuric manifestations of a desert—steppe belt
Artemisia frigida and Stipa capillata highly accumulate Cr, V
and Mo. Spiraea hypericifolia accumulates these elements in
lesser degree.In polymetallic manifestations of a steppe belt
Spiraea hypericifolia concentrates Ag, Pb, Cu and Zn, while
Artemisia frigida accumulates these elements in smaller quantities.

In gold—manifestation areas of a dry—steppe belt Spiraea hypericifolia accumulates Cu, Zn, Pb, Cr, in addition
to Mo; and Artemisia frigida accumulates V, Mo, Cr, i.e.,
aniogenic microelements.

Highest BCh—aniogenic and BCh—catiogenic activities among t
studied humidly cationic plant species are characteristic for
green mosses and among the aridly anionic and transient specie
for such plants as Artemisia frigida,Caragana frutex, etc. Lowest

aniogenic and BCh-catiogenic activities belong to humidly cationic species, such as Lonicera altaica, Sambucus sibirica, Viburnum opulus and some representatives of various grasses.

Highest BCh-aniogenic and BCh-catiogenic activities manifest in the extreme landscape and geochemical conditions, i.e., in ore deposits locations, and, most probably, on the border of species ranges, which fact we revealed in interchangeable species of the same genus: Spiraea media and Spiraea hypericifolia, Caragana arborescens and Caragana frutex, Lonicera altaica and Lonicera tatarica, etc.

BANNIKOVA I.A.
(USSR, Moscow)
ON THE NATURE OF FORESTS IN SOUTH-EASTERN KHANGAI

Many specific features are inherent in the forests of Khangai due to their being situated in a mountainous country with a sharply-continental arid climate. These forests have an insular character. They grow in the most humid habitats on the northern slopes of mountains, at considerable altitudes (1700 - 2500 m). They are characterized by a diversity of associations, expressed on rather small areas, mosaic structure of lower layers, and an abundance of psychrophilous (tundra-taiga and alpine-meadow) elements in their flora.

The route-field and stationary study of the forest vegetation in the south-eastern part of the Khangai mountain land (the Suvraga ridge, mountain systems in the upper reaches of Orkhon and in the interfluve of the Urd-Tamryn and Tsetserleg Rivers) allowed to determine the close relationship between the development of forests and specificity of hydrothermal conditions on the northern slopes of mountains. The sharply differentiated distribution of heat, inflowing to the slopes of the northern and southern expositions, conditions the wide development of permafrost-cryogenic processes in soil-ground of noninsolated slopes, with which the winter conservation of ground moisture, in particular, is connected, and, as a conse-

quence of this, additional water supply of soils in the spring-
summer period. The insular character of the permafrost processes determines the strict ecological localization of forests,
which form together with the spatially predominant steppes, a
peculiar "expositional" mountainous forest-steppe. In the
altitudinal belts of the northern slopes which differ in their
heat budgets, the processes of freezing, thawing and warming
of frozen soils and grounds take place with different intensity
determining the height differentiations of the forest belt. Its
greatest manifestation is the existence of subbelts of forest
vegetation, characterized by a definite set of forest associations, specific ecobiomorphic and floristic composition of
forests, the peculiarities of the micro-climatic environment and the direction of the soilforming processes. Within
the limits of the forest belt of South-Eastern Khangai, the
following subbelts are to be distinguished:

1. Subtaiga subbelt of grass larch forests on meadow-
forest deeply frozen soils (1700 - 2000 m above sea-level).

2. Lower taiga subbelt of grass and low bush-grass larch
and Siberian cedar (Pinus sibirica) - larch forests on turf
taiga deeply frozen soils (2000 - 2200 m above sea-level).

3. Upper taiga subbelt of moss-grass and moss-low bush
larch, Siberian cedar-larch and Siberian cedar forests on
frozen taiga soils (2200 - 2500 m above sea-level).

The grass forests of the subtaiga subbelt (Lariceta her-
bosa) are composed exclusively of Siberian larch (Larix sibirica). They develop in the lower part of mountainous forest
massives, on the border with mountainous meadow-steppes and
meadows. They are relatively open light forests characterized
by the predominance in the cover of meadow, meadow-forest and
meadow-steppe species with wide Euro-Siberian and Siberian
areas. The greatest heat budget and periodical overhumidifica-
tion of soils by permafrost waters are inherent in the habitat
of forests in this subbelt, as well as considerable fluctuations of moisture in soils and grounds in space connected with
the change of hydrothermal characteristics of ecotopes, depending on the meso-relief and orientation of slopes. Three major

types of subtaiga larch forests may be distinguished within the limits of this subbelt: "meadowed" grass forests on meadow-forest deeply frozen gleyed soils with prevalence in the cover of hygromesophyllous and mesophyllous meadow-forest and meadow grasses; "typical" grass larch forests on meadow-forest deeply frozen ordinary soils with the predominance in the cover of mesophyllous forest and marginal forbs and large-leaved grasses; "stepped" larch forests on meadow-forest deeply frozen steppized soils with a wide development of xeromesophyllous forest-steppe, meadow-steppe and mountain-steppe grasses in the cover.

The forests of the lower taiga subbelt are made up of Larix sibirica with participation of Pinus sibirica, they develop in the middle parts of mountain forest massives, and are more closed and much denser; besides species of the meadow-forest complexes, small taiga grasses and Vaccinium vitis-idaea participate in the cover.

Medium heat budgets are inherent in the forest habitats in this subbelt, also inherent are rather stable and moderate soil humidity, permafrost preservation of greater duration in their profile. During the vegetative period the soils do not become warmer than $8 - 10^{\circ}$. The major types of forests in the subbelt are Lariceta et Pineto-Lariceta pyroloso-herbosa et Vaccinioso-herbosa.

The upper taiga subbelt occupies plots on the northern mountain slopes near the upper forest border. Forest associations are formed of Siberian larch and Siberian cedar, and are characterized by being the most closed and of the greatest age (up to 250 - 300 years), and by increased windfall connected with surface development of root systems. The predominant role in the cover of forests belongs to tundra-taiga undershrub Vaccinium vitis-idaea L.,mesophyllous,taiga grass Calamagrostis obtusata Trin., and green mosses (Laricetum et Pineto-Laricetum hylocomioso-vacciniosum et bryoso-herbosum). At the upper forest border, Empetrum nigrum L. (tundra-bog undershrub), Festuca altaica Trin. (alpine-meadow grass), species of genera Juniperus, Betula rotundifolia Spach. (tundra birch)

and other psychrophilous alpine forms participate considerably
in the formation of the plant cover. The forest habitats in
the subbelt are characterized by weak heat budgets, greatly
expressed permafrost-cryogenic processes, lengthy preservation of permafrost in the upper horizons of soils and their
low temperatures $(6 - 8^{\circ})$.

Bunting B.T.

(CANADA, Hamilton)

SOIL DEVELOPMENT IN THE BRUNISOL-PODZOL TRANSITION OF THE SPRUCE-CARIBOU-LICHEN ECOSYSTEM AT THOR LAKE DISTRICT OF MACKENZIE, N.W.T., $(107^{\circ}W, 61^{\circ}N)$, CANADA

The area forms part of the little-studied subarctic black
spruce ecosystem, or forest-tundra transition, of south-central
Mackenzie district, sited on drumlinoid relief composed of
granitic sandy debris. It lies 300 km south-west of the present tree-line and is an important over-wintering area for
barren-ground caribou (Rangifer arcticus). It experiences
frequent summer forest fires related to storm activity along
the Arctic front. It has many lakes which are frozen from October to May, and has 250 mm of annual rainfall, 1000 mm snow
and only 90 mm rainfall and c. 200 mm actual evaporation in
summer (Hare and Hay, 1971). Narrow peaty areas occur adjacent
to the lakes, with Cryic Fibrisols (Pergelic Cryofibrists),
with permafrost at 15 - 30 cm depth during summer. Most of the
drumlinoid areas have Degraded Dystric Brunisols (Orthodic
Cryochrepts) on 90 per cent of the land area; while fossil
ice-wedge polygons, thaw depressions (Black, 1954) and other
limited moist sites show podzolic soils - either Orthic Humo-
Ferric Podzols, Gleyed Humo-Ferric, and Gleyed Humic, Podzols
(Day and Lajoie, 1973). In addition, frequent microsites, c. 1 m
 in diameter,   show fossil 'frost boil' soils, related to
extrusions of compacted unaltered parent material, with vesi-
cular structure, within 10 cm of the surface.

The major form of vegetation is black spruce (Picea mari-
ana) - with some Jack Pine and tamarack (Larix laricina). De-

ciduous trees are present — white birch (Betula papyrifera)
and trembling aspen. The ground vegetation apart from few labrador tea (Ledum groenlandicum) and Vaccinium vitis-idaea, is
lichen-dominated (Cladonia sp.), with lichen-heath covering
most crestal areas recently burnt or recovering from fire. The
lichen is highly-susceptible to burning, as are heaths and
labrador tea, which burning promotes rapid mineralization of
organic matter. Though soil nutrients are rapidly available
after fire, this is short-lived and slope erosion, soil instability, increased surface stoniness, thinning of organic layers and diminution of mesofaunal activity, all inhibit ecosystem elaboration — lack of biomass growth and meagre recolo-
nization by animal and bird populations. Fire effect on organic soil horizons is very variable, as is illustrated by soil
microphotographs.

From computer mapping of the soil survey, the effect of
fire may be deduced to have decreased the mean depth (4 -
10 cm) of organic layers in forest areas with age > 200 years,
to less than 5 cm, and appreciable changes result from this
burning, such as increased soil bulk density (0.2 to 0.5 gm.cc),
decrease of C/N ratios, lowered moisture saturation percentages, higher pH and lower CEC, ≤ 10 meq/100 gm, compared to
60 - 70 meq in unburned forest layers. There is very little
organic horizon regeneration (≤ 1 cm) within the first 80
years after fire, only after 200 years free of fire do soil
organic horizons attain 4 - 5 cm depth, though few forest
areas have ages greater than this.

The mineral soils conform either to the Canadian concept of
Brunisol, with thin, acid, Ae horizons of the Degraded Dystric
Brunisol, which has thin, weakly iron-illuviated, cambic B horizons (≤5 cm thick); or else the Humo-Ferric Podzols with
strong illuviation of iron (>0.6% pyrophosphate-extractable
Al+Fe). In both groups the downward mixing of organic material
is relatively weak, though some P is complexed with Fe. Organic C contents in Bf$_j$ horizons are less than 1% in Brunisols,
with bulk densities of 1.4 to 1.55 gm/cc, pH (in KCl) of 4.1
to 4.7 and low CEC (≤ 5 meq 100 gm). Free iron contents, even

in podzols, are less than 1%; total sesquioxides 2.6 to 3.0%.
In the early summer, the redox potential of the upper Bfe horizon is very low, Eh of c. 10 mv, increasing to c. 200 mv on
drying in late summer. The C horizons are usually fine to medium loamy sands with median particle size of 0.11 mm. Compact,
with bulk densities of 1.55 to 1.8, pH (KCl) c. 5.0, and Eh of
<15 mv, these C horizons exhibit coherent silt skins as re-
lict features of permafrost.

Degrees of weathering of the material in A, B and C horizons have been estimated from clay mineral and geochemical
analyses, using the standard cell approach (Beckmann, 1975).
Very localized gleyed humo-ferric podzols reveal maximum geochemical weathering, with appreciable removal of mobile ions
and greatest alteration within the solum of minerals which are
stable in the C horizons at 80 cm depth. Iron-magnesium minerals are most intensely altered in initial phases of weathering and Ca, Na and K are removed from the A horizons of podzols. Fe, Al and P are the primary depositional cations in the
B horizon and Si is the residual cation in all A horizons,
even in Brunisols. Na and K remain in the weakly-weathered A
horizons of the Brunisols.

Principal component analysis of the soil analyses showed
that the first component was of a chemical nature with the
"additive" iron and the mobilized cations revealing the fundamental differences of Ae, Bf, BC and C horizons. The second
component reflected the parent material and separated stable
from unstable weathering horizons. Together these accounted
for 77% of soil variation at all sites. From Standard Cell
Data, it was evident that Brunisols possess zones of cation
removal <6 cm thick, and accumulation and transitions horizons extend only to 12 - 15 cm. Mg is lost from the solum, Ca
from the A horizon. Fe, Al and P show gains in the Bm and
losses from the Ae horizon; Na, K and Si reveal losses in the
Bm and gains in the A horizons, the Si gain from weak podzolization, the Na and K from fire effect. In podzols, P, Mg, Fe
and Al are lost from the A, and show gains in the B, horizon,
other bases are lost from the solum. Morphologically, the

weathering zone of maximum cation instability extends to 10 cm, the accumulation zone to 21 cm and the transition zone to 25 – 33 cm depth.

X-ray analysis of the fine fractions ($<$ 20 u) shows that podzols have vermiculite and in some instances montmorillonite or mixed layer minerals, especially in Bfh horizons. Bfe horizons show 4.21 and 14.2Å peaks in many instances, related to finely-divided quartz and to chlorite-vermiculite minerals.

Thus, the difference between Brunisols and Podzol in the sub-Arctic environment depends on the relative mobility of a few components released from the most readily-weatherable minerals, as well as to the location of Podzols in localized moist micro-depressions which accumulate organic matter and are less fire-prone. Thus the research indicates that fire is an important influence diminishing podzolization and that the zonal form of soil in the subarctic forest-tundra is a Brunisol or Inceptisol and not a Podzol as is shown in most regional soil maps.

Barykina V.V., Skriabina A.A.
(USSR, Moscow, Kirov)

THE RESOURCES OF WILD FRUIT-BERRY PLANTS IN THE USSR, THEIR UTILIZATION AND PROTECTION

The Soviet Union possesses considerable resources of wild food plants. With regard to nutrients content, wild plants are more diverse than cultivated ones. Among wild plants, wild berry plants are of the greatest value. The plants of various life forms belong to this class (fruits of many of them not corresponding to botanical definition "berry"). This large group of wild plants running to more than 40 species is united with regard to these plants utilization as food resources.

In the USSR the greater part of this group species is geographically connected with north and temperate latitudes. Representatives of this group species are found from Baltic region to Far East region. Each natural zone and district is characterized by its own "set" of useful food plants which are most widely spread and are most important from the economic point of view.

They are: Rubus chamaemorus L., Empetrum nigrum L., Rubus arcticus L., Vaccinium uliginosum L. and V. vitis-idaea L., for tundra and forest-tundra. Rubus chamaemorus is very popular with the people. Empetrum nigrum is stocked by local inhabitant for winter with the purpose of personal consumption. Rubus arcticus is an interesting object for selection. The territories of taiga (boreal coniferous forest) have various conditions of wild berry growing in different geographical districts. Vacciniu myrtillus (bilberries) and V. vitis-idaea L. (red bilberries) are typical of forest watershed districts. Oxycoccus palustris Pers. (cranberries) and Vaccinium uliginosum L. (blue-bilberries are characteristic of swamps, in the northern part of the country Rubus chamaemorus is also found. Along rivers and streams Padus racemosa (birdcherries), Ribes nigrum and R. rubr and Lonicera Sp. (honey-suckle) are widely spread. Hyppophaé rhamnoides (sea-buckthorn) is peculiar of Altai and Trans-Baikal regions and Tuva. One can see raspberries, ashberries and wild strawberries in edges of forests and burns; as for the Far East, this region is characterized by the presence of Actinidia kolomikta Maxim. and Schizandra chinensis (Turcz.) Baill.

Fruit trees and fruit bushes: apple-trees (Malus silvestris L.), pear-trees (Pyrus communis L.), quince (Cydonia oblonga Mill) and alycha (Prunus divaricata Ledeb) are of great importance in the highlands of the Caucasus and in the Middle Asia.

Cranberries, red bilberries (cowberries), bilberries and blue-bilberries being the representatives of the family of Vacciniaceae L. are of the greatest importance when stocking berries and wild plants fruits. Their area occupies more than 13 million hectares. Their gross crop is more than 6

million tons and crop accessible for gathering is about 3 million tons. The Soviet Union is the main exporter of cranberries in Europe. Their stocking varies with regard to years and in some years even reaches 40,000 tons. 50% of this amount is gathered in the territory of RSFSR. According approximate calculations the territory on which cranberries are growing occupies 20 million hectares; their possible biological harvest is more than 1.5 million tons. Red bilberries take the second place in berries stocking. They have broad continuous area, which occupies more than 185,000 hectares. Their biological harvest may reach more than 2.5 thousand tons. Bumper-crop years are more often repeated in the middle taiga subzone of the European part of the USSR, in Siberia and in the Far East. Bilberries are also characteristic of coniferous forests and have rather broad area. Mainly, they are stocked in the European part of the Soviet Union. Area occupied by bilberries numbers 16,000 hectares and their biological harvest exceeds 1.5 million tons. Blue-bilberries are typical of wet places and have broad area, which spreads from Byelorussia to the Far East and occupies more than 10 million hectares. Their possible harvest is nearly 400,000 tons.

The greater part of berries crops is stocked by the inhabitants with the purpose of personal consumption; and only berries bearing continuous keeping (cranberries, red bilberries and others) are given to storing places.

Species of the family of Vacciniaceae L. occupy broad areas in the USSR, which include several natural zones and subzones.

The intensity of fructification is an integral index of species reaction to conditions of their life, so it is unequal in various parts of area. Western and central districts of forest zone in the European part of the USSR, where under the impact of Atlantic air masses the role of berries in forest communities structure is increasing, are the most favourable for berries fructification.

In rational utilization of wild foodberries some consecutive stages have been already outlined. They are: a) making inventory of resources with the utilization of the most representative methods of stock registration and mapping, b) intensification of berries harvesting with the application of modern machinery, c) making wild berries thickets cultivated, d) making berries and wild fruit plants cultivated.

Intensification of forestry and increasing the impact of recreational activity on the thickets of wild berries plants, make us pay attention to thickets protection. The protection may be considered in two aspects: 1) as prevention from infavourable natural factors impact and 2) as prevention from impact of human activities. In the first case sparseness of tracks causes many troubles in their protection (opposition to frosts, droughts and fires). Cultivation of wild thickets and their condensation (thus increasing the productivity of territory) appear more effective here. This provides increasing possibilities of thickets protection: creation of smokescreens, peat-bogs watering, struggle with plant pests, etc. In the second case, thickets protection should be realized both from human activities directed towards economic utilization of territory and from indirect human impact (forestry, recreational activity, etc.). In the case of particularly intensive recreational utilization of territory, one can notice decreasing biological potential of berries fructification. In some cases it even causes the death of their thickets. As a rule, such phenomena are observed in the districts, situated not far from populated centres.

When berries stocking is well organized, one can avoid overindustrialization, taking use of special transport means with the purpose of conveying collectors to the most remote gathering districts. In some places, teams of workers secure definite plots. In the   cases of   overindustrialization it is necessary to create temporary refuges favouring thickets restoration. The time of berries harvest should be strictly observed. Latest works show, that in some cases it is more advantageous to preserve swampy plots with berries than to use them for wood production. It is necessary to look for ways of increasing harvest volume per unit of area. One of these ways is cultivation of wild berries thickets. Some experience in this sphere has been accumulated in the USSR. However, neither wild nor cultivated thickets can't guarantee their annual rich harvest. Steady harvests may be gathered   only on condition that berries are cultivated. In some countries

the lands which do not suit other agricultural crops are used
with this purpose and definite success is achieved in this
field. In the Soviet Union such work is carried out in Estonia
and Lithuania and Western Siberia. From this point of view
cultivation of Rubus arcticus, Rubus chamaemorus, Vaccinium
myrtillus and other species is perspective, because the natural resources of these berries will constantly decrease in
connection with intensive human economic activities.

Within the USSR 3 types of territories are outlined. They
differ with regard to utilization of berries resources.

1. Sparsely populated territories, situated far from industrial centres, possessing considerable resources of berry
plants. Here intensive berry stocking is carried out, and berry protection is reduced to keeping strictly to fire-preven-
tion measures and to the time of berry gathering.

2. More populated and developed areas, where berries demand doubled attention. Here sanitary felling, thinning and
cleaning plots of forest under berries are necessary, sometimes it is advised to sow and to plant berries, to fertilize
the soil, to smoke the berries in the case of spring frosts.
One should create refuges with the purpose of preserving genetic funds of rare plants.

3. Territories, where in connection with intensification
of forestry it is profitable to create special berry plantations
(for example cranberries, sweet-briar). Soils unfit for agricultural purposes can in the nearest future become the source of
rich harvests of berries.

Byazrov L.G.
(USSR, Moscow)
SOME FACTORS WHICH INFLUENCE THE SUPPLY OF THE MASS OF EPI-
GENOUS MACROLICHENS IN THE MOUNTAIN-STEPPE ASSOCIATIONS OF
KHANGAI (MPR)

Phytomass makes it possible to evaluate quantitatively
the significance of one or another group of plants in the association, and the degree of its development. The interest

towards lichens from the point of view of determining their role in coenoses is not incidental, as these peculiar organisms draw specific substances into biological turnover.

The investigation of lichens in the south-east of the Khangai mountain-land, on the northern macro-slope of the axial ridge in the mountain-forest-stepped belt made it possible to receive information as well about the supply of the epigenous macrolichens mass in a row of grass communities in the district (See the Table).

The altitude of the locality is 1400 - 2500 m. The slopes and tops of mountains alternate with intermontane valleys of different width. Forests are concentrated on the slopes of the northern exposition at altitudes above 1700 m. The other elements of the relief are covered with steppe and meadow biogeocoenoses. It is in these latter, at altitudes of 1650 - 1900 m that the supply of the mass of above-soil macrolichens was determined 1[x]) Parmelia vagans Nyl., 2) Cladonia - C. pocillum (Ach.) O.J. Rich., C. pyxidata (L.) Hoffm., C. ochrochlora Flk., 3) Peltigera - P. lepidophora (Nyl.) Vain., P. erumpens (Tayl.) Vain., P. ponojensis Gyeln., P. malacea (Ach.) Funck., P. rufescens (Weiss.) Humb., 4) Stereocaulon sp.

Our data may be referred to the following types of habitats: V - bottoms of intermontane valleys and depressions and lower parts of slopes with frozen-bog-meadow, deep meadow alluvial and chernozem-like soils; N - mountain and sopka slopes of the northern exposition with averagehumus little-carbonaceous chernozem, meadow-chernozem, and contact-meadow soils; S - mountain and sopka slopes of the southern exposition with medium-deep dark chestnut and primitive shallow chestnut soils; T - flat tops of mountains and sopkas with primitive skeletal eroded soils. Several types of associations usually correspond to each habitat. In them, the presence is noticeable of Stipa baicalensis, S. sibirica, Helictotrichon schel-

---

[x] - numerical and letter designations are analogical to the designations on the Table.

The supply of macrolichens in some grass associations in Eastern Khangai

Habitat	Community	Cover, %		Absolute dry weight of lichens, kg/h				
		Vascular	Lichens	1ˣ	2	3	4	Total
V	Grass – sedge – forb	75	–	–	–	–	–	–
	Sedge – grass – forb	70	0.6	3.0	7.0	–	–	10.0
	Sedge – forb – feather grass	55	0.8	5.4	0.1	0.1	0.1	5.7
	Feather-grass – forb	50	1.3	2.9	2.7	–	–	5.6
	Cobriesia – grass – forb	75	–	–	–	–	–	–
N	Sedge – tipchack – feather-grass	70	–	–	–	–	–	–
	Sedge – tipchack – feather-grass – forb	50	0.1	–	0.8	–	–	0.8
	Sedge – forb – feather-grass	40	0.6	1.6	10.4	–	–	12.0
	Sedge – feather-grass – forb	50	7.1	20.6	150.0	0.1	0.1	170.8
T	Sedge – grass – petrophylous forb	35	4.3	34.8	41.2	3.2	0.5	79.7
S	Sedge – petrophylous forb – tipchack	40	2.1	31.8	3.6	0.1	2.5	38.0
	Petrophylous forb – sedge – tipchack	40	1.2	5.4	0.1	0.2	0.1	5.8
	Grass – forb	40	1.7	44.3	2.4	0.1	–	46.8
	Sedge – forb – grass	40	4.7	3.2	2.1	0.6	–	5.9
	Sedge – legumes – forb – grass	45	0.7	1.8	7.3	0.1	–	9.2

ˣdecipherment of letter and numerical designations in text

lianum, Festuca lenensis, F. pseudovina, Koeleria gracilis, out of sedges Carex korshinskyi, C. pediformis, out of forbs - Artemisia commutata, A. frigida, Stellera chamaejasme, Bupleurum scorzonerifolium, Chamaerhodos altaica, Potentilla nudicaulis, Aconitum barbatum, Geranium pratense, Valeriana officinalis, Sanguisorba officinalis, Oxytropis myriophylla, Trifolium lupinaster, Thymus gobicus, Androsace incana and others. The consecutive arrangement of associations in the Table is analogous to their position on the profile, moving from north to south, from the valley to the sopka.

The volume of the mass of above-soil lichens varies noticeably depending on the position of the coenose in the relief. The significance of separate species in the total phytomass of lichens changes, too. On the northern slope, 86 - 100% of the phytomass are made of Cladonia; while on the southern slope, more than half is formed by Parmelia vagans, leaved lichen, wich is not attached to the substratum, and is quite unnoticeable in the lowgrass steppe and mountain-steppe associations.

The tendency of lichens to settle in places with a more sparse cover of vascular plants has been noted. Two-thirds of the cases of absence of lichens on the sample areas are connected with those areas where the coverage of vascular plants made up more than 60%.

The direct dependency has been established between the cover of lichens on the substratum and their phytomass (coefficient of correlation for separate species constitutes 0.7 - 0.8).

Dyrenkov S.A.
(URSS, Leningrad)

UNE METHODE DE DIVISION D'UN TERRITOIRE EN REGIONS
FORESTIERES EN UTILISANT DES PROCEDES MATHEMATIQUES
(AVEC REFERENCE PARTICULIERE AU NORD-OUEST DE LA RSFSR)

La différenciation spatiale de la couverture biogéocoenotique de la Terre est considérée, dans le cas d'une division en régions forestières, du point de vue de l'influence

d'un ensemble d'éléments naturels et historiques sur la végétation forestière et, en premier lieu, sur la composition
et la production des peuplements. La division d'un territoire
en régions forestières représente un mode particulier de la
classification géographique des systèmes écologiques naturels. On ne saurait distinguer d'une façon stricte les dé-
cisions prises lors de la division en régions d'avec celles
d'une classification par types de forêt. Une analyse systé-
matique des unités de la classification hiérarchique générale des systèmes écologiques forestiers (tant "locaux" que
"régionaux", en termes de l'Europe moyenne) est toujours
liée aussi bien à leur structure qu'à leur essence fonctionnelle. Sous un aspect méthodologique, et surtout dans des
études ayant pour but la constitution des modèles de systèmes écologiques de rangs différents, une importance considérable est attachée au volume physique et à la forme du système
écologique considéré comme une "découpe tridimensionnelle"
faite dans la biosphère. Ce sont ces derniers qui déterminent l'ensemble de caractéristiques des composants d'un système écologique nécessaires et suffisant pour établir une
classification et l'échelle des cartes qui constituent la
meilleure des solutions connues de l'ordination écologique
des unités de classification. Il convient de savoir diffé-
rencier les unités de classification, à tous niveaux de la
classification hiérarchique des systèmes écologiques forestiers, sur la base de critères objectifs exprimés mathémati-
quement qui reflètent une série de restrictions extérieures
apportées par l'homme dans des buts pratiques. Il ne faut
pas se nourrir d'illusions sur la possibilité et la nécessité de classifications entièrement "naturelles" qui ne tien-
draient pas compte desdites restrictions.

Nous proposons ici une méthode de division des unités
géographiques administratives du Nord-Ouest de la RSFSR en
régions forestières. Selon cette méthode, on choisit des unités biogéocoenologiques suffisamment grandes (les parties du
territoire sur la carte), principalement sur la base de différences climatiques locales et géologo-géomorphologiques en

utilisant des procédés mathématiques. Ces régions forestières
ou "associations de systèmes écologiques forestiers", selon
notre classification hiérarchique, satisfont, vu le degré de
fractionnement du territoire et leur homogénéité interne, aux
conditions de l'unification de grands objectifs sylvicoles,
de la planification et, en partie, de l'élaboration de mesures
à prendre au niveau des unités géographiques administratives.
Notre méthode assure un certain rapprochement au rapport pratiquement optimal entre l'homogénéité interne des régions forestières et un fractionnement minimum nécessaire pour la
division à condition de respecter les restrictions extérieures suivantes.

1. La superficie de la région forestière doit être com-
mensurable avec la superficie moyenne du territoire de l'en-
treprise principale (en RSFSR, c'est leskhoz - économie forestière) et, en tout cas, ne pas être inférieure. Le but d'une
telle restriction est d'éviter une divergence d'échelles
entre les études scientifiques et leurs applications pratiques.

2. Les régions forestières comparées deux à deux doivent présenter l'affinité la moins possible compte tenu de la
totalité des critères utilisés pour établir les limites (contours) des régions, c'est-à-dire pour mettre au point la
classification géographique de systèmes écologiques d'un
rang donné.

3. Le degré adopté de fractionnement du territoire doit
révéler une différence pratiquement considérable et statis-
tiquement démontrable des types analogues de forêt dans deux
régions forestières voisines (les types de forêt étant les
unités d'un rang inférieur dans la classification des systèmes écologiques caractérisées par les stations identiques et
manifestant une ressemblance notable aux parties différentes
du territoire) du point de vue de la production potentielle
des peuplements et autres données sylvicoles importantes. Si
certains types analogues de forêt se trouvent identiques
dans des régions forestières voisines, c'est maintenant une
différence essentielle des régions quant à l'ensemble et à la

disposition mutuelle ("mosaïque") des types de forêt dans la
structure du paysage qui devient une condition nécessaire
pour individualiser ces régions. On rassemble ensuite plusieurs régions forestières en une "partie de l'arrondissement
forestier" en partant du niveau commun de la production des
peuplements déterminé par les conditions climatiques. Le
sens pratique de la mise à part des parties d'arronissements
forestiers consiste à établir les limites d'application d'un
ou plusieurs tableaux des types de forêt et des tables de
taxation correspondantes sur le territoire de l'unité administrative à diviser.

La méthode proprement dite comprend les étapes suivantes.

I. Etablissement d'un croquis de contours des régions
par la superposition de cartes caractéristiques particulières sur un support commun. On utilise ici des cartes hypsomé-
triques, géomorphologiques, pédologiques, climatiques, ainsi
que celles de paysage, de dépôts quaternaires et de végétation contemporaine contenant ceux des caractéristiques des
systèmes écologiques eux-mêmes et de leur milieu environnant
qui peuvent avoir, selon l'auteur, un caractère déterminant.
Aux points où il y a coïncidence d'un nombre donné (ou supérieur) de limites, est tracé le contour de la région indivi-
dualisée.

2. Examen critique du croquis avec estimation quantitative de la ressemblance/différence des régions à l'aide d'une
méthode empruntée à la théorie des ensembles. On utilise
alors les légendes de toutes les cartes caractéristiques
prises lors de la superposition. Autrement dit, on a pour
chaque couple de régions, des coefficients de ressemblance
relative pour un nombre "n" de critères.

3. Vérification des contours et leur nouvel examen afin
de les rendre satisfaisants aux critères adoptés.

4. Estimation du niveau de production des forêts déterminé par les conditions climatiques de chaque région en em-
ployant les méthodes de l'analyse des composantes principales (programme "FALES-I") et de l'analyse de corrélation -

régression (programme "PRA-3") de données météorologiques et
de leurs liaisons avec les données pour la production des
peuplements, la programmation étant faite pour la calcula-
trice "Minsk-22".

5. Rassemblement des régions forestières présentant un
niveau commun de production et des parties d'arrondissements
forrestiers; établissement de résumés des données et d'aperçus sur les conditions naturelles et historiques de chaque région.

Fraser D.A.
(CANADA, Montreal, Que.)
TOTAL GROWTH OF BLACK SPRUCE IN THE JAMES BAY AREA OF NORTHERN
QUEBEC

Tree growth was analysed according to three types of summations (Fig. 1) to indicate progressive (Fig. 2), environmental (Fig. 3) and physiological (Fig. 4) changes. Trees
were sampled from areas subject to future man-made increase
or decrease in water table levels. The techniques of this type
of tree growth investigation will be discussed in context with
past (Gaertner 1964) and future (Budyko 1974, Bryson 1974)
natural changes in eco- and macro-climates.

Fedorova N.M.
(USSR, Moscow)
GEOCRYOLOGICAL CONDITIONS AND LANDSCAPE OF THE WESTERN SIBERIA
TAIGA ZONE

The processes of formation and degradation of permafrost
were very extensive in the past within the vast area of northern and middle taiga of the Konda-Sosva Priobje area (Baulin
et al., 1967).   At present this area is characterized by
the relic insular deep-lying permafrost and contemporary
"small islands" of perennially frozen grounds, occurring near
the surface and subjected to the processes of aggradation and

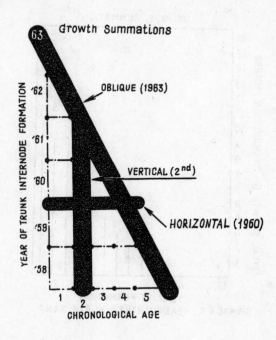

Fig. 1. Three types of summations of apical growth
of a 6-year old black spruce tree

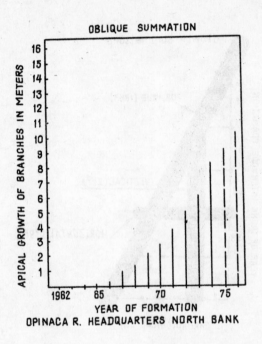

Fig. 2. Oblique summation of apical growth for a
12-year old black spruce for 1962-74 with
projection for 1975 and 1976

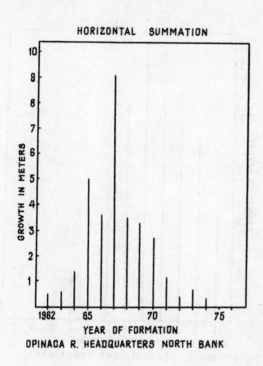

Fig. 3. Horizontal summation of apical growth for
the same tree

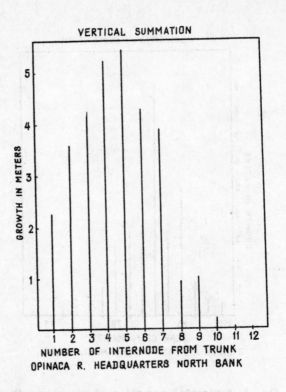

Fig. 4. Vertical summation of apical growth for
the same tree

122

degradation. Seasonal freezing and thawing of the top layers proceed everywhere. Seasonally and perennially frozen grounds under conditions of high moisture content and soil temperatures about 0° are the zone of intensive microcryoturbations (Fedorova, 1974).

These cryogenic phenomena of high regional specificity influence profoundly the soil genesis and the whole landscape and should be taken into account for the agricultural development of the area.

Perennially frozen grounds promote to a certain extent frigidity and seasonal freezing of the top layers. Relic and contemporary cryogenic and postcryogenic phenomena affect the distribution of certain land forms ranging from some tens cm up to some tens meters in length and from some tens cm to some meters in height. These forms are of various configuration including numerous closed and semiclosed depressions accumulating water and giving rise to the local bogging.

Intensive and large by mass phase conversions of moisture (water ⇌ ice) in the top layers affect in the specific way the transfer of moisture and heat, the transformations and transport of organic and mineral substances within the solum and landscape, physical, chemical and biological properties of soils, their productivity, soil morphology, soil and vegetation pattern. Migration of free and weakly-retained moisture caused by forces of gravitation, evaporation and desuction and the participation of this moisture in the macro-cycle of water (in the system ocean-atmosphere-land) is limited; a specific retention, "conservation" of moisture in soils under conditions of alternating phase state (water ⇌ ice) takes place.

Intensive microprocesses of preferentially upward migration of weakly retained nonfreezed moisture are active.

A considerable proportion of heat energy supplied to soils and occurring there is used on the phase moisture conversions water ⇌ ice (and therefore on the processes of cryogenic destruction and mass transfer) rather than on intensive and deep heating of soils. This is why the temperature of soils under conditions of moderately-continental climate stands near 0° during the most part of the year.

Biological and biochemical processes, hydrolysis of alumosilicates, illuviation and leaching of substances released during weathering and pedogenesis are also limited. At the same time the microprocesses of cryogenic local mass transfer and redistribution (including moisture, mineral particles, some nonsilicate compounds) are intensive and active to great depth in the solum.

Seasonal freezing prevents the downward migration of thaw water and promotes its transport down the slope into depressions (and the bog-formation processes in depressions). In this way seasonal freezing favours the spatial differentiation of the area according to regimes and properties of biogeocoenoses and their specific interrelations. Thus the area is characterized by mosaic (multicontour) type of soil and vegetation cover (and biogeocoenoses in the whole). Transport of soluble substances from autonomous to heteronomous soils (and biogeocoenoses) is negligible, but regimes and properties of heteronomous soils (and biogeocoenoses) are intimately connected with those of autonomous soils (and biogeocoenoses) due to the superficial redistribution of thaw water. The closely and specifically connected combination of biogeocoenoses "forest-bog" is a typical component of landscapes. Each biogeocoenosis in this combination is characterized by the regional specific features in regimes, properties and morphology. Under conditions of soil temperatures which stand near 0° during the most part of the year even small changes in climate cause considerable changes in the duration of periods when soils are in freezed or thawed state (within the limits of certain features of soils climate). These changes are especially great in heteronomous soils. Therefore the position of frozen layer boundaries are unstable in the perennial cycle (in both vertical and horizontal directions); the boundaries of soil-climatic horizons and areas (and areas of biogeocoenoses) are changeable; zones of transition between these horizons and areas are very extensive.

It should be suggested that the great similarity in geocryological conditions and hydrothermic regimes of soils in the northern and middle taiga subzones gives rise to similar principal features of pedogenesis, soil and vegetation cover (and biogeocoenoses) and to weakly pronounced nondistinct subzonal boundaries.

Examination of paleogeographical data (including paleocryological data) leads to the conclusion that during the period from the Late Pleistocene up to the modern time the geographical conditions in the area under consideration remain relatively unchanged, in the main features (frigidity of soils, their freezing and thawing against a background of high moisture content, low intensity of energy and mass macroexchange, low thickness of the layer active in hydrothermic and biological relation, cryohydromorphic appearance of the landscape).

Elaboration of methods for the development and utilization of the area should be based on regional actual data and calculations as the local specific features of the natural conditions should be taken into account. It is particularly necessary to take into account possible considerable changes in a number of parameters of the soil regimes (and related properties of landscapes) even under the weak influence of human activity.

Glasov M.V., Tishkov A.A.
(USSR, Moscow)
FUNCTIONAL PECULIARITIES OF SOUTHERN TAIGA SPRUCE FOREST ECOSYSTEMS

Complex investigations of the structure and functioning of the natural ecosystems have been developed intensively in recent time along with the traditional biogeographical approaches. The study of ecosystem as a whole, and the knowledge of the participation of separate groups of biota in its construction gives the material valid both for theory and practice.

In this study we analysed the functional-trophic structure of animal population and some aspects of the functioning of the

spruce forest ecosystems of the southern taiga. We consider
that the structural similarity of ecosystems yet does not allow to judge about typological and genetic similarity of the
communities, and only utilization of data on interrelations
between the biota components may facilitate the solving of
tasks of ecosystem typology and genesis. Spruce-forest ecosystems of the southern taiga are related by their similar high
levels of productivity, and usually by the weak development
of the super-soil cover, and by the similarity in response to
anthropogenic transformation.

The overall phytomass stock in the spruce-forests of various types is 200 - 400 t/ha with 90 - 95% belonging to spruce.
These data are similar to those, obtained for the spruce-forest on the southern border of their range, as well as for
the other type forests, for example, oak-woods. Taking into
account the statement of the zoomass zonal change and its correlation with phytomass, it is easy to discover the peculiaritie
and the rules governing interrelations of autotrophic and heter
trophic links of ecosystems in different geographic zones.

The most important factor for the animal inhabitants of
spruce-forest is the trophic one, i.e., the quantity of the
available phytomass. The spruce-forests produce from 6 - 8 t/ha
per year, which is close to the corresponding data for forest-
steppe (11 - 14 t/ha). However, the biggest part of the production falls on the lignine (more than 60%), which does not
participate in the heterotrophic cycle in stand formation.
92 - 93% of the spruce-forest biomass are unavailable for
phytophagous species. Corresponding data are somewhat lower
for other forest ecosystems of the subzone and the other zones
(for the oak-forest of forest-steppe, for example).

Thus we may consider the spruce-forest ecosystems as de-
tritus type ones. The specific features of the animal population of the southern taiga are the low values of the overall zoomass (for instance, 5 - 10 times lower than that in
the forest-steppe), the limited number of species with sharp
density fluctuations, and the stenophagia of the main fauna
components. The structure-functional analysis of spruce-forest

ecosystems showed the full domination of saprophagous species
for both density and zoomass, while correlation between phytophagous and predators zoomass may change with seasons to the
advantage of either group. The increase of the predators fraction is an evidence of the tension of interecosystem connections, fulfilled by this group (partly by the consuming of the
fraction of the "air plankton").

The scale of animal utilization of spruce seeds and seedl-
ings with significant frequency of the years rich in seeds
(2 - 3 years), puts the animal influence on the spruce repro-
ductive cycle in the class of the leading processes in functioning of spruce-forest ecosystems. On the basis of observations in spruce-forest in Novgorod region and the comparison
of the obtained data with those for the other analogous locations, we shall consider some patterns of the influence of
zoocomponents on spruce-forest ecosystem functioning in southern taiga.

The phytomass production in spruce-forests is quite uniform from year to year, with the exception of climatically ano-
malous years. The main production of the green mass takes place
during a short period (2 - 4 weeks). It is that time when the
phytophagous species impact is most significant. The overall
invertebrates zoomass of the tree crown can reach more than
10 kg/ha. Up to 50% of the annual shoots, mainly of the upper
or middle parts of the crown, is injured by invertebrates (and
to a less extent by vertebrates). During the whole vegetation
period, predators dominate (60%), among which the spiders are
most numerous (90%).They play the main role in the regulation of
both active phytophagous forms and air plankton density (mainly Diptera). The activity of vertebrate phytophagous species
in utilization of the current spruce-production is not
significant and can be very pronounced only in the years of
low-seed-yields (the damage of buds and shoots by squirrels
and crossbills).

The litter input in spruce-forest is observed all the year
round and most intensely - in summer (2 - 4 t/ha). The organic
mass stock in the litter is considerable (10 - 30 t/ha, depen-

ding on the type of the spruce-forest and its age). For the
forest-steppe oak-forest these values are much lower, what can
be explained by the rate of decay and by the abundance of sapro-
trophic organisms in soil and litter of the oak-forest. The
heterotrophic inhabitants of the litter in spruce-forest are
presented by bacteria, fungi, yeast and litter-soil fauna of
invertebrates. The density of bacteria in soil under the spruce-
forests is high - 500-1800 thousands ind.per 1 gr of soil,the
density of fungi - up to   100 fungal embroys per 1 gr
of soil. Among fungi the dominants are Penicillium and Morti-
erella. The leading role in decomposition of the vegetation
remnants in spruce-forest belongs to bacterial, fungal and
yeast flora. That differs them from ecosystems of deciduous
forests. The important role of invertebrates in the process
of decomposition is shown for oak-forests of forest-steppe.
They decompose more than 70% of the litter, and the micro-
flora transforms only 9%. The big differences may be explained taking into account that the litter and the upper layer of
the soil in spruce-forests are very poor in invertebrates
$(0.8 - 1.0 \text{ g/m}^2)$. The main biomass is presented here by saprophagous (Lumbricidae, Diptera larvae), by predators (Aranea,
Carabidae, Stafilinidae, Chelopoda) and to the less extent - by
phytophagous organisms (Elateridae).

Among microartropods Collembola and Oribatei dominate
(70 - 80% of all the group). The correspondence between the
large stock of dead organic matter and small mass of decompo-
sers could be explained on the basis of peculiarities of litter composition and chemistry. The role of invertebrates in
decomposition of the windfallen spruce trees is more important than in the litter and upper layers of soil. The inverte-
brate biomass in stumps and logs is 20 times more than in the
litter, but the species composition at the last stages of the
decomposition is similar to that of soil-litter.

The seed yields in the southern taiga repeat approximately
every 3 years. However the more frequent yields are possible.
For example in Valdai upland the yield was observed in 1973

and 1975. The whole cones   yield in 1973 was damaged by cater-
pillars of Laspeyresia strobilella.

65 - 70% of full-grain seeds were excluded by all animal
groups (vertebrates and invertebrates) before their flying
out and germination. The rest seeds germinated and added to
the soil supply. The animals do not only use seeds as food but
also take part in the seed redistribution.

As discussed above, the other types of spruce-forests of
the southern taiga show the same regularities. Ecosystems where
edificators are the species of wide geographic distribution
possess ecological plasticity and similar character of organization. Comparative uniformity of structure, dynamics and processes of metabolism under changing geographical conditions is
achieved in such ecosystems by similarity of phytoclimatic characteristics and sufficiently stable composition of the edificators consortions. Such qualities are characteristic for the
"ripe" communities with fastened adaptive properties of the
components. The spruce-forests ecosystems in southern taiga
may be certainly referred to the "ripe" communities.

Radvanyi Jean
(France)

LE ROLE DES RESERVES DE HAUTE MONTAGNE DANS LES ALPES ET LE
CAUCASE

La haute montagne, dans les pays tempérés, est un lieu
idéal pour implanter des parcs ou réserves. Paysages naturels,
espèces animales et végétales ont depuis longtemps incité
l'hómme à délimiter des zones où la nature serait protégée et
présentée aux touristes. La première créée dans les Alpes
françaises (parc du Pelvoux) date de 1913; la réserve de Lago-
dekhi dans le Caucase de 1912. Cependant la comparaison du
fonctionnement des parcs créés depuis et leur rôle fait apparaître à côté de problèmes communs, des différences sensibles.
Cela tiént aux conditións antérieures à la création du parc
(densité de l'occupation humaine surtout) et au statut juri-

dique, avec et c'est fondamental, l'application dans la pratique des statuts. Nous prendrons en exemple en France le
parc de la Vanoise (60.000 ha de 1000 à 3852 m créé en 1963)
et celui des Ecrins (Parc du Pelvoux élargi, 91.700 ha de 800
à 42102 m, créé en 1973); en U.R.S.S., la Réserve du Caucase
(266.000 ha de 1200 à 3300 m créée en 1924) et celle de Téberda (83.400 ha de 1500 à 4042 m, créée en 1936). Dans sa liste
des parcs et réserves analogues de 1967, l'O.N.U. classait
quatre réserves du Caucase (les trois citées et celle de Rit-
za) et prenait en France le Parc de la Vanoise (seul créé à
cette date) avec des réserves dues aux statuts des parcs
Français qui ne correspondent pas aux critères de l'O.N.U.
Les réserves du Caucase, quoique dépendant d'organismes différents (ce qui est un des aspects négatifs du système soviétique), ont un statut à peu près analogue: créées par décret
du Conseil des ministres de Géorgie ou de R.S.F.S.R, elles
ont un territoire vaste, dont la quasi-totalité fonctionne
en réserve intégrale. Une partie seulement est accessible aux
touristes (à Téberda, 100 à 300 ha sur 83.400). Les statuts
type, dès les années vingt prévoyaient un triple but: conser-
ver la nature, l'étudier, aider au développement du tourisme
alors très limité. Toute activité économique (même le patu-
rage), la chasse, la pêche, la cueillette sont strictement
interdites. La direction comprend une grande part de forestiers et de scientifiques. Il y a un laboratoire permanent
dans chaque réserve. Par rapport à ce type qui entre tout à
fait dans les critères de l'O.N.U., les statuts des Parcs Nationaux français présentent deux originalités importantes:
à cause de la densité de la population et de l'ancienneté de
l'occupation, le législateur a expressément prévu la continuation (même contrôlée) des activités rurales, essentiellement
l'estivage. Les réserves intégrales sont très rares. De plus
on prévoit pour des motifs surtout économiques une "zone pé-
riphérique" ou "pré-parc" qui entoure le parc lui-même et
doit connaître un important développement économique, en principe contrôlé, avec des restrictions (chasse interdite). Cette
zone est plus grande que le parc: en Vanoise 143.700 ha;

pour les Ecrins 178.000 ha. Le parc a donc comme buts: conser-
ver la nature (mais dans son état humanisé - d'après certains
textes officiels, le paysan est le gardien et le protecteur
de la nature et on doit subventionner une agriculture non ren-
table mais nécessaire à la conservation du paysage), organiser
le tourisme, provoquer l'essor économique de régions en crise.
Nous voyons apparaître deux positions opposées. Les Soviéti-
ques dans les réserves du Caucase laissent les équilibres naturels se rétablir. La limite de la forêt remonte, les anciens
prés de fauche se transforment en fourrés que l'on voit déjà
passer à un pré-bois. La forêt elle-même reprend son caractère
naturel, les troncs pourissant sur place, ce qui ne manque
pas de majesté. Le problème paysan ne se pose pas; il est ré-
solu indépendamment des réserves. A l'opposé, la position
française, dictée par les conditions pratiques (les rapports
avec les paysans sont très difficiles) recherche curieusement
une justification théorique: il serait absurde du point de vue
biologique de laisser la nature sans influence humaine; le
nombre d'espèces diminuerait; la nature laissée à elle-même
est laide... Un autre problème est le respect des statuts et
la vie des parcs, surtout en fonction du tourisme. De part et
d'autre, on cherche à le développer ce qui revient souvent à
mettre en cause les statuts. Les exemples les plus nets sont
le projet d'hyper-station de ski en Vanoise et le projet de
route à travers la Réserve du Caucase (tous deux critiqués
et rejetés). Comment rendre compatibles protection et tourisme? Dans l'optique français, c'est possible si on reporte les
infrastructures lourdes dans le pré-parc (en évitant l'anar-
chie urbanistique actuelle) et si on se limite dans le parc
aux randonnées pédestres ou à ski, safariphoto, etc. Dans
l'optique soviétique, c'est plus délicat. Il est difficile de
concilier réserve intégrale et développement du tourisme. Le
ski alpin apporte sans cesse de nouvelles nuisances (problème des nouvelles pistes à Téberda-Dombaï). De là l'idée de
certains scientifiques: créer des sortes de parcs où l'on mê-
lerait protection et tourisme et des réserves destinées à la
recherche et au "tourisme scientifique".

Dans tous les cas, il nous semble indispensable d'effectuer des études écologiques, phytosociologiques précises pour apprécier après quelques années de fonctionnement les modifications apparues.

Miah M.M.
(BANGLADESH, Dacca)

EXFOLIATION OF SANDSTONE IN A TROPICAL ENVIRONMENT: AN EXAMPLE FROM BANGLADESH

Weathering of sandstone in an island situation in Bangladesh is being investigated. Huge blocks of rocks lying in three different physiographic situations - the intertidal zone, an inland marsh, and a higher and relatively drier ground - show different forms of weathering. Examination shows that exfoliation may be caused due to one or more of the following factors:

          a) moisture,
          b) temperature,
          c) internal structure.

Moskalenko N.G., Tagounova L.N.
(URSS, Moscou)

LA DYNAMIQUE ANTHROPOGENE DE LA VEGETATION DE LA SIBERIE DU NORD EN RAPPORT AVEC LES CONDITIONS ECOLOGIQUES

A l'heure actuelle la création des constructions linéaires de grandes dimensions a pris une large ampleur à l'occasion de la mise en valeur des gisements de pétrole et de gaz dans les régions plates du nord caractérisées par la congélation des sols perpétuelle. L'élimination ou la déformation de la végétation dans ces régions provoquent des modifications considérables dans l'environnement (renforcement ou affaiblissement des processus exogènes, variation du régime hydrothermique dans les sols, de la profondeur de leur congélation ou décongélation saisonnière, etc.). Toutes ces

modifications peuvent avoir des conséquences indésirables
(par exemple, la déformation des constructions). L'étude des
variations de la végétation dues à l'influence technogène et
la prévision de leur dynamique ultérieure présentent donc un
intérêt pratique, mais elles ne sont pas encore suffisamment
argumentées.

La prévision de la dynamique anthropogène de la couverture végétative est basée sur la connaissance des lois de sa
dynamique naturelle, l'étude des variations typiques de la
végétation sous l'influence technogène, la mise au point des
tendances de son développement, la comparaison avec la dynamique des phytocénoses modifiées par la construction en conditions analogiques.

Les travaux expérimentaux pour l'établissement de telle
prévision ont été effectués en Sibérie du Nord.

Les recherches de la dynamique naturelle de la végétation de la Sibérie effectuées par B.N.Gorodkov (1946) et
A.P.Tyrtikov (1974), ainsi que nos propres recherches de dix
ans sur le terrain et de cinq ans au stationnaire des variations de la végétation sur le tracé des constructions linéaires, sont à la base de la prévision mentionnée.

L'étude de la dynamique des phytocénoses a été accompa-
gnée par les mesures du régime hydrothermique des sols et des
roches pédogénétiques, de la dynamique de leur décongélation
saisonnière, par les observations microclimatiques de la température, de l'humidité de l'air et de la vitesse du vent sur
les terrains permanents situés dans les associations végétales les plus répandues. Ces terrains ont été choisis en conditions naturelles, ainsi qu'en conditions modifiées.

Ces recherches ont permis de caractériser les variations
dans les phytocénoses sous l'influence des constructions linéaires et de mettre au point les tendances de leur développement ultérieur. Grâce à l'utilisation des photographies
aériennes de grande échelle faites avant et après la construction on a pu porter toutes les modifications sur les cartes.

Pour la vérification des tendances évidentes du développement des phytocénoses et pour leur extrapolation à long

terme, on a étudié les tracés anciens. On a fait la comparaison des résultats de l'interprétation des photographies aériennes de grande échelle faites avant la construction et 15 ans après. L'analyse de cette documentation permet de déterminer le caractère et le degré de modification de la végétation et d'évaluer la cadence de sa restauration. La photo-interprétation a été suivie de l'étude de l'état contemporain (23 ans après la construction) de la couverture végétale et de sol, ainsi que de l'état de congélation sur les profils traversant le tracé des constructions linéaires.

Sous l'influence technogène la couverture végétale est soumise aux variations considérables en fonction de différentes conditions de l'environnement. Cependant, après l'intervention unique dans la situation naturelle la restauration progressive des phytocénoses est encore possible.

La cadence de restauration du tapis végétal est différente et dépend des complexes naturels se distinguant par le relief, les sols, les conditions d'humidification et géocriologiques. L'influence la plus négative sur la restauration du tapis végétal est exercée par la modification de la couverture du sol et l'insuffisance de l'humidité.

En fonction de la cadence de restauration du tapis végétal on peut grouper les complexes naturels d'une façon suivante: bandes et cavités allongées de l'écoulement avec les marais à linaigrettes - à laîches - à sphaignes sur les sols tourbeux; terrains plats marécageux raréfiés de mélézin à sous-arbrisseaux et à sphaignes sur les sols tourbeux à gley avec des lentilles des sols constamment gelés; terrains à petits mamelons avec les associations à sous-arbrisseaux - à sphaignes - à lichens sur les sols tourbeux podzoliques illuvials-ferrugineux avec des lentilles des sols constamment gelés; tourbières plates à sous-arbrisseaux et à lichens sur les sols tourbeux gelés; terrains drainés plats de lignes de partage des eaux avec les bois clairsemés à cèdre, à pin et à mélézin à sous-arbrisseaux - à lichens sur les sols sableux podzoliques dégelés; mamelons tourbeux minéraux et bourrelets à cèdre et à sous-arbrisseaux et à lichens sur les sols gelés tourbeux-podzoliques et illuvials-ferrugineux sableux.

Ainsi, sur les bandes et cavités allongées de l'écoulement la restauration du tapis végétal se passe plus vite. Vers le commencement de la cinquième année après l'intervention technogène les groupements à linaigrettes et à laîches et ceux à linaigrettes- à laîches - à mousses se forment et couvrent 80-100% du sol. 20 ans après, les phytocénoses développées sur le tracé des constructions linéaires traversant les marais ne se diffèrent pratiquement pas des associations végétales naturelles.

Par contre, sur les mamelons minéraux les conditions pour la restauration du tapis végétal sont moins favorables à cause de la destruction partielle du sol, de l'insuffisance de l'humidité et de l'action du vent. Quatre ans après les variations des conditions on peut rencontrer ici des groupes isolés de laîches (Carex globularis L) et des tâches rares des mousses (Polytrichum strictum Sm) ne couvrant que 10-20% de surface. Sur les mamelons qui ne sont pas fixés par la végétation l'affaissement du toit des sols pergelisés et le développement des processus éoliens peuvent avoir lieu. Pendant la construction il faut éviter ces terrains ou bien il faut prendre des mesures pour la formation artificielle du tapis végétal qui peut assurer la protection contre le développement des processus d'érosion et thermokarstiques.

Les données obtenues sur la dynamique anthropogène du tapis végéral ont été utilisées pour l'établissement des cartes prévisionnelles de paysage et d'indication appliquées à l'établissement de la précision des conditions géocriologiques. A la base de l'établissement de ces cartes sont les cartes des complexes naturels et les données sur le rapport du tapis végétal avec les autres éléments du paysage observés aussi bien en conditions naturelles qu'en conditions modifiées.

La prévision de variations des conditions géocriologiques dues à la création ou à l'exploitation des différentes constructions doit être basée sur l'analyse de la dynamique et de l'évolution des complexes naturels dans la situation

naturelle ou sous l'influence des facteurs anthropogènes, y
compris l'analyse de la dynamique du tapis végétal, composant le plus instable et difficile à renouveler.

E.M. Naumov
(USSR, Moscow)
THE SOIL GEOGRAPHY OF THE NORTH-EAST ASIAN TAIGA

The soils of the Extreme North-East of Asia have been poor
investigated until recently. There were no data reliable concerning the geographical regularities of their distribution.
In the 40-50's all the soils of the North-East taiga were supposed to belong to podsols. Then, in 60's, podsols were considered to be the characteristic soil type for the region along
the Ohotsk Sea shore only, and absolutely absent in the continental territories of Yana, Indigirka and Kolyma. As a result
of our soil-geographical investigations the basic types of
taiga soils were determined and the main geographical regularities of their distribution found. These regularities are
shown at the soil map presented here.

As it was found the inner differentiation of the soil profiles and the soil mantle of the region is attributed to the
peculiarities of the local factors of the soil formation,
while the severe climate and the north taiga vegetation are
common for the whole territory. The investigations made it
possible to divide the taiga territory into two soil-geographical provinces: the humid one, occupying the Ohotsk Sea
shore and the continental one, in the Yana-Kolyma region which
differ sharply by the presence of differentiated and non-differentiated profiles in the soil mantle. As it was founded for
the Ohotsk Sea shore province, soils with well differentiated
profile are widely spread there. This fact is caused by the
humidity of climate in the first hand and the light composition of the weathering crust due to predominance of granitoids.
The predomination of the podsol humus-illuvial soils (Al-Fe

humus) is the characteristic feature of the territories along
the sea shore. The specific genus of the ochrohumus-illuvial
soils occupying considerable territories was distinguished
among them. These soils are developed on the acid volcanic
glasses. The area of their distribution along the Ohotsk Sea
shore is broaden to the east and narrowed to the west. Soils
with non-differentiated profile have limited distribution and
occupy minor positions in the province. They are located at
the flat loamy territories (mainly boggy ones) and the northern slopes on the shallow permafrost or on the long-seasonal
frozen ground.

The gleyic podsols on the permafrost and the boggy cryogenic soils are widely spread at the flat territories - river
terraces and gentle slopes. As far as the continental Yana-
Kolyma province is concerned, it should be mentioned that the
soils with non-differentiated profile occupy the major territories of the soil mantle there. This fact is attributed to
the continental type of climate and besides it to the predominance of the shale rocks and the clay loam weathering crusts.
Non-podsolic humus-illuvial soils (podburs) were founded to
be the mostly distributed on the stony crusts and merzlotozems (cryozems) on the clay loams. Vast territories are covered
with the boggy cryogenic soils.

In defiance of the routine opinion that there could be no
podsols in the continental provinces the possibility of podsolization under the favourable combination of soil formation
factors has been revealed. Podsols are located in the peripheral parts of the province and the subtundra areas of the
mountain-taiga belt notable for being more humid than the inner plains and the closed intermountain depressions. The formation of podsols on the rocks which do not delay the eluvial-
illuvial process and favour relative quartz accumulation in
the bleached horizon at the free drainage in profile, is the
other condition of podsolization.

The most continental semiarid and arid areas of the Yana-
Kolyma province (Oimyakon-Verkhoyansk) are characterized by
the specific combinations of the unique pale-cryogenic satura-

ted soils, taiga-steppe/dry-steppe chernozem-like and chestnut-
like/soils and podburs. These soils occupy the warm south
slopes of the mountains and form the specific combinations
with merzlotozems developed at the slopes with the cold exposition. The pale taiga (cryoarid saturated) soils are developed under the xerophytic larch-woods of the north taiga at the
non-calcareous low-stony loamy eluvial-deluvial deposits.

The reaction close to neutral, the complete saturation
of the whole mineral profile (while the litter is acid), absence of the excessive moistening, verkhovodka and gley formation process and the presence of the dense cryogenic horizon
at the bottom of the profile - are the characteristic features
of these soils. At the southern slopes the cryoarid soils form
combinations with the taiga-steppe (dry-steppe) soils, developed under the sheeps fescue-wormwood-feather-grass associations on the stony-loamy eluvial-deluvial non-calcareous deposits. The profile of these soils is characterized by presence
of the upper sod-humic accumulative and calcareous accumulative
horizons at the depth of 20 - 50 cm.

The soil mantle of the mountain-tundra belt territories
which are situated higher than the mountain taiga is characterized by combinations of the tundra podburs, tundra cryogenic gley soils and outcrops of rocks.

We suppose the regularities of the soil geographical distribution founded to be typical for the taiga territories of
Alaska and the North Canada as well.

Nosova L.M., Stavrova N.I.
(U.R.S.S., Moscou)

PARTICULARITES DE L'EXTENSION GEOGRAPHIQUE ET DE LA
LOCALISATION ECOLOGIQUE ET PHYTOCENOTIQUE DE L'OXALIS
ACETOSELLA L. ET DU CAREX PILOSA SCOP

Une étude des aires sur le fond des processus naturels
permet de résoudre beaucoup de problèmes concernant la nature
écologique, l'origine des plantes, lés particularités floris-
tiques et géographiques de certaines régions etc. Avec cela

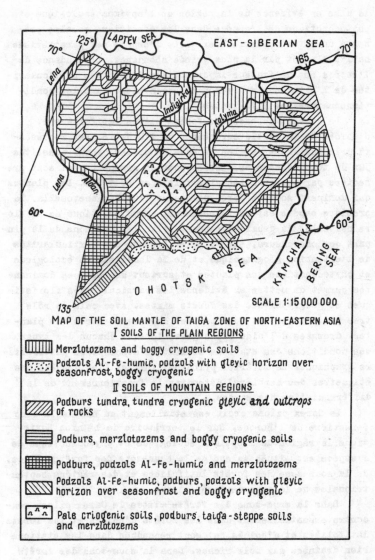

MAP OF THE SOIL MANTLE OF TAIGA ZONE OF NORTH-EASTERN ASIA

I SOILS OF THE PLAIN REGIONS

- ▥ Merzlotozems and boggy cryogenic soils
- ▦ Podzols Al-Fe-humic, podzols with gleyic horizon over seasonfrost, boggy cryogenic

II SOILS OF MOUNTAIN REGIONS

- ▨ Podburs tundra, tundra cryogenic *gleyic and outcrops* of rocks
- ▤ Podburs, merzlotozems and boggy cryogenic soils
- ▦ Podburs, podzols Al-Fe-humic and merzlotozems
- ▧ Podzols Al-Fe-humic, podburs, podzols with gleyic horizon over seasonfrost and boggy cryogenic
- Pale criogenic soils, podburs, taiga-steppe soils and merzlotozems

SCALE 1:15 000 000

I39

la mise en évidence de la région de l'optimum écologique et
phytocénotique de l'espèce dans les conditions zonales présente un intérêt particulier. Ici, l'espèce est caractérisée
habituellement par la plus grande abondance. L'abondance de
l'espèce est un facteur important témoignant des particularités dé l'existence et de son rôle cénotique dans les conditions naturelles différentes.

Le couvert herbacé des forêts mixtes latifoliées et
d'épicéa de la partie européenne de l'U.R.S.S. est représenté principalement par les espèces némorales ce qui constitue
59% du total d'espèces. Les espèces du groupe boréal sont présentées par une quantité plus faible et font 25%. Les plantes
qui dominent sont le Carex pilosa et l'Oxalis acetosella. La
première espèce est un représentant caractéristique de la flore némorale, la deuxième espèce, selon les opignons de la plupart des chercheurs, est boréale. L'étude des particularités
de l'extension géographique et de la localisation écologique
et phytocénotique des plantes et surtout des espèces dominan-
tes permet de mettre en évidence les particularités floristi-
ques et géographiques des forêts mixtes. Avec cela le rôle
très particulier est joué par les cartes des aires des plantes, dressées à l'aide des signes, lorsque chacun des types
des conditions des stations obtient la description particuliè-
re graphique. Le matériel principal initial (les données des
étiquettes des herbiers) détermine le fractionnement de la
différenciation des signes.

Le Carex pilosa croit essentiellement au sud de la zone
forestière de l'Europe. Sur le territoire de l'Union Soviétique la région de l'optimum écologique et phytocénotique de
l'espèce est située au sud de la sous-zone des forêts mixtes,
de la sous-zone des forêts latifoliées et des régions septentrionales de la zone de forêt-steppe.

Dans la sous-zone des forêts mixtes le Carex pilosa rencontre essentiellement dans les forêts d'épicéa et les forêts
latifoliées et d'épicéa qui sont répandues dans les stations
bien drainées aux sols riches. Dans la sous-zone des forêts
latifoliées le Carex pilosa, au contraire, se rencontre sur

les sols plus pauvres, en cédant à l'Aegopodium podagraria
aux stations avec les sols riches. Les forêts de chêne et celles de chêne et de tilleul sont les phytocénoses les plus caractéristiques dans la composition desquels dans cette sous-
-zone se rencontre le Carex pilosa. Cette espèce est la plus
répandue dans la zone de forêt-steppe dans les régions des
plateaux Volyno-Podolskoïé, Srednie Rousskoïé et Privoljskoïé
où elle se rencontre en abondance sous le couvert des forêts
dé chêne mélangées avec le charme, l'érable, l'orme, le tilleul.

Vers le Nord de la région de l'optimum écologique et phytocénotique le Carex pilosa se rencontre très rarement et seulement dans les régions confinant à la sous-zone des forêts
mixtes. Dans la région des steppes des graminées et des her-
bes différentes on peut rencontrer le Carex pilosa dans les
conditions non zonales, c'est-à-dire dans les forêts latifoliées insulaires.

L'aire de l'Oxalis acetosella est vaste. La partie essentielle de l'aire couvre presque tout le territoire de l'Europe, l'exception fait le Nord extrême et le Sud de même que les
étendues considérables de la Sibérie. Cette espèce est aussi
connue en Extrême-Orient, au Japon, en Corée, à l'Hymalaya et
en Amérique du Nord. Aux résultats de l'étude on a été constaté que dans les conditions de Plax les sous-zones de la taïga
méridionale et des forêts latifoliées et d'épicéa sont les
plus favorables pour la croissance de l'espèce donnée où son
rôle comme le dominant du couvert herbacé dés phytocénoses
forestiers est exprimé dans une grande mesure. Cela permet de
considérer les régions géographiques mentionnées en tant que
les régions de l'optimum écologique et phytocénotique de l'espèce donnée.

Au sein de la région de l'optimum l'oxalyde se rencontre
dans les conditions phytocénotiques différentes. Le plus souvent cette espèce joue le rôle du dominant du couvert herbacé
des forêts sombres résineuses du type de la taïga méridionale
et des forêts d'épicéa composées avec une grande présence
dans le coúvert, le sous-bois et souvent dans le peuplement

forestier, des représentants de la flore némorale. Dans la
sous-zone des forêts mixtes de la partie européenne de
l'U.R.S.S. L'Oxalis acetosella est en abondance sous le couvert des forêts mixtes composées de Picea abies, Quercus ro-
bur, Tilia córdata, etc.

Vers le Nord de la région de l'optimum écologique et phytocénotique, dans la sous-zone de la taïga moyenne l'Oxalis
acetosella se rencontre essentiellement sur les sols les plus
riches et bien drainés des pentes sud et dans les vallées. A
la proximité des frontières septentrionales de l'aire dans la
taïga septentrionale de la partie européenne de l'U.R.S.S. et
de la Sibérie occidentale l'Oxalis acetosella devient une espèce très rare.

Vers le Sud de la région de l'optimum écologique et phytocénotique, dans la sous-zone des forêts latifoliées le rôle
de l'oxalide diminue aussi considérablement. Ce n'est que dans
les régions occidentales de la sous-zone elle se trouve souvent parmi les dominants du couvert herbacé des forêts composées de hêtre, de chêne, de frêne et d'aune glutineux ce qui
permet d'incorporer ces régions dans la région de l'optimum
de l'espèce.

Aux Karpates et au Caucase cette espèce est représentée
largement dans le couvert herbacé tant dans les forêts sombres
res résineuses d'épicéa et de spain comme dans les forêts latifoliées de hêtres. Dans la zone supérieure forestière
l'oxalyde est en abondance dans les forêts mixtes avec la présence de hêtre et des essences sombres résineuses. Sur les
montagnes de l'Oural, de Saïane, de Kouznetsky Alataou
l'Oxalis acetosella se rencontre principallement dans les
forêts sombres résineuses d'épicéa et de spain.

Dans les forêts latifoliées et résineuses et les forêts
latifoliées de montagne de l'Extrême-Orient le rôle de l'oxalide en tant que le dominant du couvert herbacé est relativement faible bien que cette espèce soit son composant perma-
nant.

Ainsi, dans le couvert herbacé des forêts mixtes latifoliées et résineuses de la partie européenne de l'U.R.S.S.

l'ensemble spécifique des espèces a été composé dans lequel
les espèces de l'élément floristique némorale prédominant
dont le Carex pilosa est d'un des représentants caractéristiques. Sur le territoire de l'Union Soviétique cette espèce
est répandue essentiellement dans les forêts latifoliées et
seulement dans la sous-zone des forêts mixtes elle se rencontre dans les phytocènoses résineux dans lesquels, quand
même, dans l'étage arborescent, le sous-bois et le couvert
hérbacé les espèces némorables sont représentées en grande
quantité.

D'après sa nature écologique, l'élément boréal, moins
représenté par son nombre d'espèces, est hétérogène. Ici, on
distingue les espèces particulières du type de l'Oxalis acetosella qui sont communes dans les forêts latifoliées et
d'épicéa les forêts latifoliées et celles de la taiga méridionale et c'est pourquoi il est plus correcte de les appeler
les espèces némorales et boréales.

P.OZENDA (GRENOBLE, France)

LES GRANDES LIGNES DE LA VEGETATION DU CAUCASE
VUES PAR UN BIOGEOGRAPHE ALPIN

Les recherches géobotaniques faites par l'Institut de
Biologie Végétale de l'Université de Grenoble dans l'ensemble
de la chaîne alpine ont permis d'établir (P.OZENDA 1966, Documents pour la Carte de la Végétation des Alpes, vol.IV):

a) que les groupements végétaux actuellement connus dans
cette chaîne, au nombre de plus de trois cents, peuvent être
répartis d'une manière satisfaisante en 32 séries dynamiques
appartenant elles-mêmes à 5 étages de végétation: méditerranéen, collinéen (y compris supraméditerranéen des Alpes du
Sud), montagnard, subalpin et alpin;

b) que les types d'étagement observés dans la végétation
des différents massifs alpins peuvent se réduire à 5 types
principaux: A_1 Préalpes du Nord-Ouest (Chartreuse, Préalpes
savoyardes); A_2 Préalpes du Nord-Est (Suisse, Bavière, Autri-

che); A$_3$ vallées internes (Briançonnais, Val d'Aoste, Valais, Vintschgau); A$_4$ Alpes subméditerranéennes (Haute Provence); A$_5$ type illyrique (Slovénie, Alpes vénètes, mais aussi en partie Alpes maritimes et ligures).

Nous avons pu montrer aussi que les séries et les types d'étagement définis dans les Alpes suffisent à représenter d'une manière approchée la végétation des chaînes voisines: Jura, Carpátes slovaques, Apennin du Nord et du Centre, Dinarides.

En vue d'éprouver la validité de ces concepts dans les chaînes extra-européennes, nous avons étudié l'Himalaya Népalais et établi l'existence de types d'étagement qui rappellent la disposition verticale observée dans les Alpes, mais naturellement avec une importante translation (1800 m environ) des étages homologues en raison de la différence de latitude entre les deux chaînes (DOBREMEZ 1972).

ZIMINA et Coll. (Revue de Géographie Alpine, 1974) ont décrit dans le Caucase 5 types d'étagement, qui sont notés ici "K":

K$_1$ Kubanien; K$_2$ Colchidien; K$_3$ Daghestanien; K$_4$ Lagodech-Zakatalien; K$_5$ Arménien (celui-ci dans le petit Caucase seulement).

Au cours d'un Colloque franco-soviétique sur la comparaison géographique entre le Caucase et les Alpes occidentales (Juillet-Août 1974), une partie de ces types ont été identifiés; il semble qu'il faut en ajouter deux autres que nous nommerions K$_6$ Baksanien et K$_7$ Centro-Géorgien (ce dernier peut-être très proche de K$_4$). De mes propres observations au cours de ce voyage, je pense pouvoir dégager les conclusions suivantes:

1°/ On peut proposer, du moins à titre d'hypothèse de travail, les comparaisons: A$_1$ - K$_1$; A$_3$ - K$_6$; A$_4$ - K$_4$ et K$_7$; A$_4$ pro parte et A$_5$ - K$_2$; K$_3$ ne semble pas avoir d'équivalent dans les Alpes, dont aucune partie n'est franchement steppique. Mais ces comparaisons sont valables essentiellement pour les étages collinéen et montagnard, moins évidentes par contre pour les étages subalpin et alpin (par exemple la grande ceinture subalpine à Larix et Pinus cembra des Alpes internes n'a pas d'équivalent dans le Caucase).

2°/ La steppisation progressive qui s'observe dans le Caucase en allant de l'Ouest, où elle ne concerne que l'étage inférieur, vers l'Est où elle est générale dans tous les étages, est dans une certaine mesure comparable à l'influence méditerranéenne dans les Alpes françaises du Sud, où les groupements subxériques ne touchent dans le Dauphiné que l'étage collinéen mais remontent en Haute Provence jusque dans l'étage subalpin qui possède des groupements à Astragales épineux.

3°/ La comparaison de tous les étages de végétation des différentes parties des deux chaînes est un travail considérable mais qui peut être simplifié en utilisant au départ la similitude entre certaines formations. Il semble recommandable de partir de l'étage montagnard dans lequel la forêt de Conifères sombres caucasienne est certainement l'équivalent écologique de la Hêtraie-Sapinière des Alpes, tandis que de leur côté les Hêtraies à Fagus orientalis du Caucase semblent diversifiées en constituant un ensemble assez parallèle à celui que forment les subdivisions des Hêtraies à Fagus silvatica, et notamment les sous-alliances du Fagion, dans les Alpes.

4°/ Une autre direction de travail qui pourrait être exploitée en priorité est la comparaison des groupements supraforestiers dans les deux chaînes, car ces groupements sont maintenant connus d'une manière très précise dans les Alpes. D'après des observations qui ont pú être faites au cours du Colloque 1974 dans la région de l'Elbrouz, les groupements prairiaux de l'étage subalpin supérieur du Caucase se distribuent écologiquement suivant un schéma qui est analogue à la mosaique des groupements prairiaux subalpins des Alpes. Par contre l'étage subalpin inférieur du Caucase est assez différent de celui des Alpes et formé surtout de brousses et de landes à Bouleaux et à Rhododendrons comme dans l'Himalaya, de sorte qu'il vaudrait mieux, pour marquer cette originalité, employer le terme de "subcaucasien" plutôt que celui de subalpin.

Rodman L.S.
(USSR, Moscow)

FLOODPLAIN MEADOWS BELOW THE DAMS ON RIVERS WITH CONTROLLED
FLOW

Controlled flow of rivers as a result of hydro-construction became of universal importance for the last 20-25 years,
which fact makes it urgent to evaluate the changes in nature
systems under continuous influence of controlled flow.

Major changes here occur in ecosystems of floodplains.
Vast areas are used under reservoirs, upstream rises of ground
waters take place and much land is swamped. Greatest damage,
however, as hydro-construction practice in Finland, Czechoslovakia, FRG, USA and USSR shows, is done to natural fodder
areas situated in the tail-waters of hydro-technical structures. Here, the levels of high waters drop, their duration
shortens, natural periods shift, low water level and temperature cycle changes, also disturbed are processes of river
channel. The zone influenced by the new hydrological conditions is rather large and ranges from 100 km to 500 km. Complicated problems that sprang up after the construction of hydro-
technical projects on the Volga, the Don, the Dnieper, the
Irtysh, and on other rivers, show how important it is to consider the after-effects of the controlled flow. Still the
question about the changes that happen in natural vegetation
of floodlands after the control of the flow is not studied
thoroughly enough. There are some works in which it is merely
ascertained that changes do happen, but the mechanism of the
influence of new conditions on vegetation is not revealed. To
solve the problem continuous observations should be conducted
in fixed areas before control of the flow is introduced and
after it.

The character and tendencies of changes in floodland vegetation depend on the geographical position as well as on the
extent of difference between artificial and natural conditions.
In humid areas decrease of the level and the duration of high
waters may positively influence the condition of natural fodder

grounds which was registered, in particular, in floodland
meadows of the Ob and the Irtysh basins (Aseeva, Ilyina,
1973). In semi-arid and arid zones the above changes lead to
unfavourable consequences. So, during the primary filling of
the Bukhtarma reservoir on the Irtysh river (1961 - 1963),
floods were not permitted into after-bays, which led to the
desiccation and resalinization of floodland soils, impoverish-
ment of meadow vegetation, wide spread of weeds and drop in
meadow yields by 5 or 6 times. At the present time in under-
waters of the Bukhtarma and the Tsimliansk hydroelectric power
plants they practice artificial floods, though irregularly,
once every four or five years which fact brings in catastroph-
ic successions - in nonflood years there can be observed the
shift to galophytization and steppe formations, while the dis-
turbance of natural herbage is accompanied by wide spread of
weeds. Each flood time brings the vegetation back to the earlier stages of the succession process (Gorbachov, Lutsenko,
1973). As a result of the filling of the Kapchagai reservoir
the development of meadow vegetation sharply deteriorated in
the present delta of the Ili river, resulting in xerophytiza-
tion and abundance of annual halophytes and weeds (Nikitev-
itch, 1974, Plisak and others, 1974).

Thus, the violation of the natural floodland conditions
in arid and semi-arid zones leads to a sharp change in natural vegetation with its yield efficiency which is 4.5 - 6 times
greater than that of watershed and they provide fodder stock
for cattle-raising in large territories. Annual spring compensation flows constitute the necessary condition for preservation of meadow vegetation in floodplains of arid zones.
In recent years spring compensation flows are practiced on
the Irtysh, while on the Volga they are carried out regularly
since the start of the construction of the Volgograd Hydroelectric Power Plant. In these cases the extent of vegetation
changes depends on corresponding accuracy of simulated floodland conditions and natural ones. It also depends on whether
the controlled flow causes the violation of the complex dynamic balance between floodland vegetation and hydrological conditions by which it is determined.

These problems were studied on the example of the Volga
Akhtuba floodplain situated in tail-waters of the Volgograd
hydrotechnical project inaugurated more than 15 years ago. In
this section the Volga does not have tributaries and the influences of the controlled flow are especially obvious. The
1958 - 1973 investigations were carried out by way of repeated observations on key sections and ecology-geobotanical profiles along the whole stretch of the Volgo-Akhtuba floodplain.

Annual man-made floods allowed the preservation of forest
and meadow vegetation in the Volgo-Akhtuba zone. However, in
the period of controlled flows the duration of floods shorten-
ed by 10 to 15 days, their time changed, while the upper part
of the floodland at its middle level experienced rare inundation and meadows of high level were nearly deprived of it. As
the result, vegetation specific for arid and steppe-type meadows extended, they shifted down the relief, the number and
abundance of weeds, as well as ephemerals and ephemerids also
increased. Along the entire stretch of the floodplain decrease
in general moistening and a certain rise in salinization of
soils can be observed. Though, on the whole, the entire floodplain retains the predominant humid-meadow moistening (according to L.G. Ramensky's scale), a shift toward steppe is registered everywhere as well as the reduction of locations with
swampy moistening. In areas with 100 percent moisture supply
where it was only the duration of inundation that reduced,
meadows humidification went down in an average by 3,1 points,
while vegetation changes were of no definite character. But in
zones completely deprived of inundation or with poor supply,
humidification went down by 5,8-7,2 points (as summarized for
the entire floodplain); desert-steppe weeds and perennial
zonal semidesert plants appeared, and changes in composition
of the dominants and associations were registered. This new
position cannot be considered stabilized since the leading
role here belongs to explerent species, weeds, and the ephemerals. If conditions do not change these would, undoubtedly,
be forced out by zonal semi-desert species, i.e. we shall wit-
ness successions registered on the Don. But the return of lo-

cations to flooded conditions, short as it might be, would surely lead to restoration of meadow vegetation.

The work that has been done brings us to the conclusion that even under secured flooded conditions changes in supply and duration of inundation lead to accumulative changes of vegetation. Good knowledge of the character, specifications (direction) and rapidity of these changes are the basis for long-range forecasting of the state of vegetation under set conditions, as well as for optimizing the latter. Detection of the initial stages of the process is also of primary importance. Accumulation of satisfactorily representative data on the remote-future consequences of the construction of hydro-technical projects in ecosystems of different natural zones is imperative.

Rudneva E.N., Tonkonogov V.D.
(USSR, Moscow)
GEOGRAPHICAL ASPECTS OF PODZOLIZATION ON SANDS

1. Podzols and podzolized soils on sands are widely spread over the area of the Eurasian continent. These soils are developed under a wide range of climatic conditions, on various (in mineralogy and texture) sandy deposits. They occur in the arctic tundra, taiga, forest-steppe and steppe, in the oceanic areas of Western Europe, in the regions of Siberia and Yakutia with most continental climate and on the Asiatic Pacific coast. Both sandy podzols and podzolized soils display a distinct eluvial-illuvial profile differentiation due to the Al-Fe humus podzolization with the formation of podzolic albic and spodic horizons.

2. The wide range of climatic conditions under which podzols and podzolized soils on sands occur, leaves a strong imprint on their properties: with the decrease of heat from the middle and northern taiga of the Russian plain with the most pronounced podzolic profiles, to the East-European tundra, for example, the depth of podzolic horizon and of the whole solum falls, while the humus illuviation increases and gleyization

over the permafrost appears. The increase of both heat and humidity (humid tropics) brings about the growth of thickness of all the horizons, especially of the humus and podzolic ones. The increase of humidity to the west of the Russian Plain is accompanied by the increase of depth of peaty or humus horizons and by the growth of humus reserves in the solum, especially in the illuvial (spodic) horizon. Greater continentality (East Siberia, Yakutia) is marked by weaker participation of the oxalate-extractable iron oxides in the formation of the illuvial R_2O_3 maximum due to their dehydration in highly contrasting climatic conditions. In sandy soils formed under more warm and arid climate (southern taiga, forest-steppe and steppe of the Russian plain) the accumulative humus horizon appears, all the genetic horizons become less pronounced, in the transitional zones between them more diffuse. The humus illuviation in the B horizon diminishes and may be practically absent (forest-steppe and steppe).

3. Properties of sandy podzols depend on mechanical and mineralogical composition of sands. The latter may be estimated by the total sesquioxides content. The increase of sesquioxides content or of the amount of rock fragments in the parent material causes the decrease of the podzolic horizon depth alongside with that of whole solum. In the same time the humus content in the B horizon grows, the share of the brown humic acids in the humus rises and the participation of non-silicate R_2O_3 in the formation of spodic horizon is reduced.

4. Mineralogical and mechanical composition of sands influences the soils sensibility to the alteration of external climate as well as the soil hydrothermal regime. In podzols on polymict sands the subzonal climatic peculiarities are distinctly visible in the soil properties. In the middle taiga, for instance, as compared to the northern one, the total depth of the solum increases, the podzolic horizon becomes thinner, directly under the forest litter an accumulative-raw-moder-humus horizon appears, the eluvial-illuvial profile differentiation of organic matter weakens.

The podzols areas on quartz sands enlarge considerably as compared to that on polymict sands, whereas the climatic changes, even the zonal ones, do not influence the soil properties. Podzols occurring in southern tundra, northern and middle taiga differ quite insignificantly. Podzols on stony or polymict sands have a more favourable thermal regime. They thaw slightly earlier and are heated stronger and deeper during the growth season. Cultivation of many crops is easier on such soils under the unfavourable climate in high latitudes.

5. Podzols and podzolized sandy soils occurring in climatic zones with different atmospheric moistening occupy various landforms because of varying ratio of precipitation and superficial run off as constituents of the total soil moisture regime. Thus, in the northern and middle taiga subzones with their most favourable conditions for podzolization the sandy podzols absolutely predominate in the soil cover structure of sandy areas. They occur as on micro-and meso-elevations and on flat landforms, so in the shallow weakly pronounced depressions, being replaced by water-logged soils only in more profound depressions. In case of excessive atmospheric moistening (typical and southern tundra) sand podzols are restricted to extremely "dry" sites - flat microelevations, while soils of shallow microdepressions display many features of boggy soils. In arctic tundra with its weak soil weathering the areas with podzolized sandy soils are rare. These soils may be formed here only under conditions of very weak additional moistening (weakly pronounced nano-depressions).

Soils of microelevations are not podzolized and soils of microdepressions may be regarded as boggy ones. Sandy podzols of the southern taiga subzone may be found only in the lowest parts of soil catenas. They usually occupy rather shallow flat depressions and the most even low parts of the slopes receiving additional moisture of the surface run off and lateral through flow. The autonomic mesomorphic sites are dominated by sod-podzolic or weakly podzolic (surface-podzolic) soils with a spodic B horizons. Under insufficient moistening conditions (forest-steppe and steppe) the podzolized sandy soils

become very rare, occurring as small patches in depressions
due to additional surface moisture.

Vitko K.R.
(USSR, Kishinev)
TO THE CONTACT ZONE PROBLEM IN GEOBOTANICAL SUBDIVISION

One of the important and insufficiently solved problems
related to geobotanical subdivision is the contact of subdivision units (regions, provinces, districts). Distinct boundaries are rare, as a rule there are more or less wide intermediate contact zones, or strips (Karamysheva, Lavrenko, Rachkovskaya, 1969), where a complicated pattern of communities
(phytocoenoses) with some characteristic features of both con-
tacting units occurs. This pattern and the composition of intermediate zone communities should be taken into consideration
when a concrete line at a map is being drawn. That is why all
kinds of communities are to be preserved in a contact zone. It
is very important to study contact zone vegetation between contrasting botanical-geographical regions. Species of such regions are characterized by different ecological peculiarities.
Species edificators at their limits are more sensitive to some
ecological factors.

The knowledge of these peculiarities is necessary both
for resolving geobotanical subdivision problems and for eluci-
dating vegetational cover history (forests genesis under climate changes and anthropogenic activity).

We deal with the problems concerning the forests of the
Moldavian Socialist Republic. Moldavia is situated at the extreme southern-western part of the USSR. The geobotanical subdivision of Moldavia has been proposed by T.S.
Heideman (1952, 1964, 1966). On the territory of the republic
there is a contact of three botanical-geographical regions
such as European broad-leaved forest (middle-european and east-
european provinces), Mediterranean forest (euxine province)
and Euroasiatic steppe (pontic province) regions. The limit
between forests with the predomination of mesophylous Quercus

petraea Liebl. and xeromorphic submediterranean forests with
Quercus pubescens Willd. is due to the specific geomorphological, climatic and soil conditions. The intermediate strip between middle-european and euxine provinces with the southern
part of Central-Moldavian highland, the so-called Codry, with
narrow watersheds (at average 200 - 300 m a.s.l.), high proportion of slope surfaces and the variety of ecological conditions. The contact zone belongs to a diffusion type (Karamysheva, Lavrenko, Rachkovskaya, 1969; Gribova, Isachenko, Karpenko, 1972). There is a range of associations differentiated
by the role of mesophilows middle-european and hemixerophilous
submediterranean species in their structure. In some communities middle-european species predominate in all strata (Quercus petraea - Crataegus curvisepala - Cartex pilosa), in
others - submediterranean species predominate in uppermost
stratum and in understorey; there are submediterranean species in herbaceous layer too, but an invasion of steppe and
ruderal species usually takes place there (Quercus pubescens -
Cotinus coggygria - Festuca valesiaca). Then there is a range
of communities where middle-european species predominate in
arborescent layer, but submediterranean species are widespread
in understorey (Quercus petraea + Cotinus coggygria + Lithospermum purpureo-caeruleum). Other combinations may occur too.
Such intermediate communities have been named buffer communities by Sotchava.

The specific forms of combinations of communities in the
contact zone are complexes made up of phytocoenosis fragments
with predomination of Quercus pubescens or Q. petraea. They
occupy rather vast territories in southern Codry. Similar complexes have been found by us in the Crimea. It means that such
complexes represent a natural kind of association in the marginal zone between forests with the predomination of Q. pubescens
and Q. petraea.

Not only plots of climax forests but some mentioned above
buffer communities and complexes as well are under protection
in Moldavia.

Vergara R. Ricardo
(Cuba, Havana)

FACTORS OF DISTRIBUTION OF CUBAN FRESH-WATER FISHES

An analysis of the main causes which have determined the extant Cuban fresh-water ichthyofauna as to its diversification and distribution of the taxa involved, is presented on a basis of interregulationships about habitat diversity, geo-historical events, periodical climatic changes, and morphological and physiological adaptations of the species considered.

Results of the above-mentioned analysis, founded on a comparison of the continental emerged areas, chiefly Central America, with the other West Indies (Trinidad excluded because of being a continental island, with an essentially South American fauna), points to: (1) Cuba is a secondary evolutionary center; (2) it has not reached a full ecological equilibrium as to saturation and overlapping of available ecological niches; and (3) the suty of the native taxa shows that all of them have two common features: eurihalinity and development of adaptations to periodical changes in physical conditions of the environment.

Because of its greater size compared with the remaining West Indies (more accentuated during some time in Quaternary, having into account the insular platforms) and its greater antiquity, as it is inferred from the prevailing present tectonics in the Caribbean Region, it could be considered that Cuba has acquired physical and temporary conditions which favour the speciation processes, and explain its greater taxonomical diversification.

Another important factor which has contributed a great deal to such diversification is its closeness to the emerged continental areas (more pronounced during eustatic changes occurred in the Quaternary). On account of this fact, there are found in Cuba taxa of nearotic origin (i.e., Lepisosteus and Fundulus), jointly - with the neotropical ones, which

are, by far, the most abundant. Nevertheless, as the closest
phylogenetic relationships are with the Central American
species, it can be concluded that the greatest part, if not
all, of the Cuban fresh-water ichthyofauna is of Central
American origin, having Yucatán as the most likely distribution center.

Though the seas which surround Cuba are very deep and are
traversed by marine currents – being both factors geographical barriers – it must be taken into account that fishes from
Cuban rivers belong to secondary fresh-water groups. All of
them are euryhaline to a variable extent and in the course
of its evolutions have developed adaptions and strategies
to their environments, though they can – achieve their vi-
abilities(i.e., the great eurihalinity of Cyprinodontidae,
the viviparity of the poeciliids and Lucifuga, the catadromy
and adaptation to torrent life of Joturus and Agonostomus,
and the aerial breathing of Lepisosteus).

Cuba, in comparison with the remaining West Indies, is
most differentiated concerning natural regions, lesser
proportion of mountains (in relation with its area) and greater extension of Karst. All these factors are in themselves
geographical barriers, which in turn favour the speciation
processes. Thus, in its territory are found exclusive taxa,
sometimes extremely localized, such as Cichlasoma ramsdeni
from the Guaso River basin; the blind-fishes of the genus
Lucifuga (with two species; recently a third species has been
described in the island of New Providence, Bahamas), the cy-
prinodontid Cubanichthys, found in a discontinued distribution, in some rivers of the western part; and most of all, the
diversified genus Girardinus, with eight species, and its
closely related Quintana.

The above-mentioned facts can establish the hypothesis
that Cuba is a secondary evolutionary center, as to fresh-
water fishes, through its the West Indian territory having
the richest ichthyofauna. Central America, because of its
greater climatical, topographical and taxonomical diversity,
is considered as a primary evolutionary and distributional

center, if compared with the Antillean Region. It is known that the first mentioned area is the center of origin of the Cyprinodontoidei; it has a greater species richness in those groups found in Cuba (excluding the auctocthonous ones) and has a moderate number of Siluriformes, South American invaders, absent in the West Indies (not having into account those species introduced by man).

Though Cuba is not, in fact, a bridge between Central America and the remaining Antilles, it can be established a gradient of taxonomical distribution, as follows: Central America (Yucatán) – Cuba– Haití– Porto Rico – Lesser Antilles. Jamaica show greater affinities in its species composition (those belonging to the genera Poecilia and Gambusia) with those of Haití, and, thus, it can be established an ichthyogeographical unity, comprising both islands. The Bahamas are very impoverished in this connection: only two Criprinodon, one Gambusia and one Rivulus.

A curious phenomenon of exclusion of mutually competitive groups is observed between Cuba and Haití: one, out of two, taxon is richer in species in one island than in the other. Thus, Poecilia, with nine species in Haití and only one in Cuba, affords the most striking example. Competition is considered as the main factor in the dispersion of this taxon, even more than the availability of ecological niches or the geographical effect. The scarce success of Poecilia in Cuba can be accounted for the presence in this island of the closely related and diversified genus Girardinus (plus Quantana).

Competition could be, also, the cause of the isolation of some taxa in the Cuban territory. Thus, the cyprinodontids are found in the geographical periphery and are most plentiful in brackish waters, where they has been displaced by the most reproductively successful poecilids, which are of more recent origin; Joturus is seldom seen in our mountain torrents because of the presence of Agonostomus, more successful, as its geographical distribution shows, both in the West Indies and in the continental areas; and Cichlasoma ramsdeni, which is circumscribed to a basin, whereas in the remaining Cuban

I56

territory is very widespread the plastic - C.tetracantha
(as is reflected in its high polymorphism).

Finally, and as a basis to consider that Cuba is not fully
stabilized in its fresh-water ichthyofauna, it must be taken
into account the presence in the rivers of: (1) incursory
marine taxa, such as Tarpon, Centropomus and Mugil, which
are very frequently found in plain fresh-water; and (2) the
introduced North American taxa, such as Micropterus and Lepo-
mis, whose success is rather moderate, as it is inferred from
the fact that, apparently, they have not produced any cata-
strophic effect in the native species, both by predation and
completition.

V. GENERAL PHYSICAL GEOGRAPHY

Drdoš J.
(ČSSR, Bratislava)

A COMPLEX PHYSICO-GEOGRAPHIC ANALYSIS OF THE NATURAL ENVIRON-
MENT DEMONSTRATED ON THE EXAMPLE OF LIPTOVSKÁ VALLEY
(WESTERN CARPATHIANS)

The last decade in the Czechoslovak geography is characterized by developing of a complex point of view, especially in the physical geography as theoretical considera-
tions (L. Mičian, 1971, J. Drdoš, 1973, 1974, J. Kvitkovič,
1973, P. Plesník, 1973, J. Demek, 1974) and complex physico-
geographic studies (E. Mazúr et al. 1971, J. Demek, 1973,
J. Drdoš, 1967 et al.).

The term complex physical geography means in our country
a study of the vertical and horizontal structure of the natural part of the geosphere, that is the system of litosphere,
pedosphere, hydrosphere, atmosphere and biosphere. A.G. Isač-
enko designates this subject epigeosphere (1965). The vertical
structure of the natural part of geosphere means the mode of
bounding of the respective geo-spheres, which is variable in
the space and time. This variability in space manifests itself
as horizontal structure. This phenomenon is conditioned by
various exogenic and endogenic factors, strongly influenced by
manifold human activities.

The study of the vertical structure of the natural part
of the geosphere, called also internal structure, and her variability creates the basis of the complex physical geography.
This study, besides the analysis of the influences of endogenic and exogenic factors and activities of man unveils the
regularity of its horizontal structure formation.

Thus the complex physical geography endeavours to identify the regularities of spatial structure formation of the natural part of the geosphere on basis of analysis of their temporal-spherical variabilities.

For the complex physico-geographical analysis in the Liptov's Valley we have chosen the territory of the River Váh plain, which is uniform with regard to genesis and present processes of modelling.

The differentiation of the territory is conditioned by a small number of factors, as the soil texture and the petrographic characteristics of the substrate, the depth of ground water level, the relief, the human activity, etc.

On the basis of the character of linkage among the natural components of the territory, namely between the substrate, the hydrogeographic relations, the soils as well as the vegetation, which reflex the properties of the abiotic components, 7 types of natural complexes and 17 subtypes have been identified.

Four types of natural complexes are directly bound with the intensive dynamic of modelling processes of the river plains, two of them are independent on it and one is antropogenous. According to the relations to the effect of the ground and surface water it is possible to distinguish among the natural complexes of the river Váh plain three dry types, the characteristics of which are not formed by the ground or surface water influence, two types are humid, strongly influenced by the forming forces of inundations and moderate to strong influence of the ground water and two types of natural complexes are a direct product of the action of ground and surface water.

The physicogeographic structure of the river Váh plain in the Liptov Valley is formed by a mosaic of the following natural complexes: A group of types of natural complexes connected with intensive modelling processes on the plain and with a different degree of ground water influence. The group is coherent to plain terraces, deposited cone accumulation and the lower degree of the plain. Characteristic are the alluvial soils (brown, also typical and gluey). Territories

with semihydromorphous soils are changed to agricultural
lands. There are partly mesophilic meadows with communities,
where Arrhenaterus elatior, Festuca pratensis, Poa pratensis,
etc. dominate. On the more humid parts with gluey soils are
Carex-communities where Carex gracilis, Carex hirta, Carex
flava, etc. prevail. The territory is divided according to the
types namely due to nonuniform effects of groundwater, which
provides soils of different properties.

Four types of natural complexes has been created: a type
of natural complexes without a remarkable influence of inundation ground water, a type of natural complexes with moderate
influence of inundation and ground water and a type of natural
complexes with strong effects of inundation and ground water.
A group of natural complexes, connected with permanent action
of ground and surface water. The group is bound to locations
of springwells of ground water, where swamps are formed. The
presence of water is a dominant factor in forming the properties of the hydromorph natural complexes. The group is divided
into two types according to the soil character. The first type
has a swampy soil with a proper vegetation, in which species
as Scirpus palustris, Typha latifolia, Phragmites communis etc
prevail. To the last group belongs the anthropogenic type of
natural complexes formed on made-up-grounds. They have a coar-
se anthropogenous soil and mezophilic meadow vegetation.

The structures of natural complexes of the river Váh plain at Liptovska Valley result from the action of different factors, as inundation, action of ground water, the substrate, etc
The natural complexes of a higher relief degree are more stabilized than complexes of the river plain, which is subject to
the strong effects of inundations and ground water, which
are, due to intensive dynamics of the modelling processes more
labile. These structures are strongly influenced by man. Anthropogenous changes can be observed on all types.

Hadač E.
(ČSSR, Prague)
LANDSCAPE AND MAN

Our Landscape went through several stages of development.
The first landscape appeared on the Earth's surface with the
first soild rock. This was the first stage, the inorganic
landscape, formed by the geosphere only. The second stage
started with the origin of life as the transformator of solar
and cosmic energy into a biogeochemical one. Life has been
modelling the face of the Earth for more than three billion
years, changing the whole atmosphere, causing chemical weathering, creating soils, producing limestone and coal deposits –
forming the Biosphere. The third stage of the Landscape development started with the appearance of Man. We can thus speak
about inorganic landscape, biotic landscape and inhabited landscape.

Mutual relations between Man and Landscape changed drama-
tically during ages, but in all times they formed a dialectic
unit, influencing each other. At first, small herds of Man did
not differ much from herds of other animals; they collected
their food or chased animals, and were chased by beasts for
prey. Being an integrate part of geobiocenoses, they did not
change the Landscape any more than other animals. But none the
less, foundations to the technosphere were laid in this period,
and mutual influencing between man and his technisphere started. In this period Man's "technique" consisted mainly of stone
tools and sometimes fire. Its radius was at insignificant, but
its influence on Man was strong enough, changing successively
his adaptive hand as well as his brain.

Radical changes in the relations between Man and Landscape started in the Neolithic period, some 10,000 years ago.
Man burned forests to prepare pastures for his herds, and his
cattle, sheep and goats made the forest regeneration on clear-
ings impossible. The same effect resulted from crop cultivation. Natural landscape changed into cultivated one. With increasing amount of meat and cereals at his disposal, Man could

defy famine and the number of men increased accordingly. At first, he was a typical predator on Landscape, recklessly taking from it everything he could have use of, destroying often valuable natural resources. Successively, he found out that it does not pay to destroy natural resources and eventually a secondary equilibrium between Man and Landscape was established, simulating the homeostasis of natural ecosystems, geobiocenoses.

Raised animals from pastures and crops from fields were consumed and all wastes went back to the fields and pastures, so that the productivity of soils was not destroyed. Thus originated a new type of ecosystems, the technoanthropocenose (TAC).

Man's technical world at this period was much more diversified if compared with the poor equipment of the palaeolithic hunter. He had a house, better formed stone tools, constant fire, earthen ware, leathern or woolen clothing, etc. During several millenia Man's technical equipment and his agrotechnical methods, his TAC did not change much. In this relatively long period close interrelations and connections were established between all the various members of this agrarian technoanthropocenose, and not only formed, but also fixed by many successive generations. Man and his domestic animals, his fields and cultivated plants grew so interdependent that they could not exist isolated any more; they formed something like a big family.

All the members of this big "family" got individual names and were respected as individuals – horses, cows, dogs, cats, even fields got their names.

Further development of TAC went generally towards a higher agglomeration, integration, specialization and differentiation following the evolutionary low of sociability, formulated by V.J. Novák. Originally a loose connection between some few agrarian TAC for protection against enemies gave eventually origin to city states. Later on, if one of the towns got some advantage over other towns, a state originated with one capital and several towns.

Cities and towns had manysided functions: protective, administrative, cultural, cultic, they concentrated artisans and tradesmen, merchants, etc. In contrast to the agrarian TAC, which by its character was "autotrophe", able to feed itself without substantial help from outside, the urban TAC is by its character "heterotrophe", dependent on the agrarian TAC in respect of food. The transition from "autotrophy" to "heterotrophy" was gradual and took centuries. At first, the towns were "mixotrophe", as most of the citizens has fields below the town walls and thus took part in the production of victuals. Later on, bigger towns were transformed in purely "heterotrophic" formations.

The relations between agrarian and urban TACs took different forms. In most cases there was a well balanced symbiosis. Agrarian TACs provided urban TACs with food and raw material and got industrial wares, profited from protection against enemies, had their cultic and cultural center in the town. There was none the less always a part of parasitism in their relation, mainly on the side of the urban TAC. Sometimes, on different levels of culture and civilization, this parasitic component prevailed, like in the classic Sparta. There, the town was a military organization, parasitising openly on agrarian TACs and giving them practically nothing.

From the medieval towns, centers of trade, administration, culture, etc. or besides those, industrial centers appear in modern times, producing an increasing quantities various more or less useful products, new machines, etc. Factories are attracting more and more people to the town and finally monstrous agglomerations originate like London, Tokyo or New York.

Simultaneously, the symbiosis of rural and urban TACs is becoming very tight. Whereas previously rural TACs were practically independent from the urban ones, nowadays they depend on industrial centers in indispensable machines and especially in energy. Whereas before the main energy used in rural (as well as in urban) TACs was Man's and horse's, in modern villages electric power and fossil fuel are indispensable. Renewable energy resources were exchanged for non renewable ones.

The transition of rural into urban TACs was, as we have
shown, a gradual one, but the modern industrialization went
much more quickly. Human biology and psychology were not able
to adapt themselves to the entirely new environment. The interrelations and connections of the rural TAC were so solid
that they persist even in this fully new industrial environment - in a changed form and function. The connection between
Man and animals in the rural TAC was based on mutual aid: the
dog was watching the house or cattle and got food for it, cat
was fed to catch mice, hen to give eggs. In town keeping of
dogs and cats has lost its original function. The dog does not
watch, the cat does not catch mice - they are so to say surro-
gates, filling the now unoccupied niche in the man's psyche.
Similarly, the landscape itself is at last transformed into a
landscape picture on the wall, giving Man's fantasy the possibility to place himself into the open landscape and for a
while to relax his nerves, tired by his industrial environment.

Man did not get used to his technosphere, on the contrary, he gets nervous and aggressive like friendly baboons, when
placed in cages, as suggested Edward Goldsmith in one of his
speaches. But on the other hand, Man is today unable to live
without his technosphere. Even Robinson could survive in wild-
erness only by building his new, though primitive technosphere.

To find a new equilibrium with the landscape, his natural
environment, Man has to simulate natural geobiocenoses, i.e.
not to destroy but to recirculate most of its natural resources,
creating as little wastes and emissions as possible.

Novák V.

(CSSR, Brno)

ON THE APPLICATION OF SOME MODES OF REPRESENTATION IN THE COMPILATION OF THE MAPS OF THE PHYSICO-GEOGRAPHICAL REGIONALIZATION OF THE CZECH SOCIALIST REPUBLIC

A part of the research work of the physico-geographical and economic-geographical regionalization of the Czech Socialist Republic are the thematic maps of this territory representing the phenomena investigated in the form of analytic, polythematic and synthetical maps.

The maps have been compiled on the scale of 1:500,000 and are related to all partial and resulting themes of both regionalizations. The given scale is very favourable for the territory of the Czech Socialist Republic since it allows to concentrate a great quantity of cartographical information in the map and to represent the whole territory of interest on one map sheet of size A 0.

The thematics of the physico-geographical regionalization is represented by 15 maps orientated to the regional and typological division of the relief, morphometry (relative relief, mean altitudes, mean angles of slope), climatic division, hydrological conditions (density of streams, territorial division of shallow groundwaters and surface waters) pedogenetic and granulometric associations, potential erosion, biogeographical division, forest density and resulting physico-geographical regions. All phenomena investigated form continua establishing the methods of area representation.

The economic-geographical regionalization involves 5 partial themes (immigration regions, commuting regions, functional classification of communities, agricultural and industrial regions) with one resulting theme (economic-geographical regions) based on the presupposition that the regions constructed are always of nodal type and their delimitation is given by boundary values. Except the maps of agricultural regions scattered detached objects are concerned which can be represented by means of the area method owing to small scale and substantial generalization.

Many of the maps mentioned represent not only an area drawing of the phenomena investigated but indicate the casuality conditioning the constructed or recorded phenomena. These maps should be paid greater attention to. Here some maps of the physico-geographical regionalization are concerned on which the resulting continual phenomena have been represented by means of areas the basic means of representation being the colours.

Among these maps belongs the map of the typological division of the relief on which the various relief forms are represented by colours so that the colour is not only of qualitative (e.g. plains, highlands, etc.) but even of quantitative significance giving by its gradation partly even a hypsometric image. The morphostructures are represented by hachures of different inclination. Each unit is characterized even by a trinomial number the first cipher designating the relief form, the second the structure and the third the mode of development which is not represented by drawing. The localization of the units is made possible by a reduced situation given in inexpressive colours. The map is an example of graphically simply represented resulting phenomena the conditional character of which is represented at least partly.

The regions of surface waters from the space mentioned are an example of a multilayered map. The constructed units – the hydrogeographical regions represented in the map by lines are characterized by specific runoff (6 colour tints), retention capacity (colour shades substituted by the width of colour stripes – 4 degrees), runoff variations (linear signature with 5 symbols) and runoff coefficients (5 degrees expressed by letters of small alphabet). The last characteristic is recorded only in a quadrinominal numeral symbol given for the various regions. Even this map has been completed with a reduced and inexpressively drawn situation. The considerable quantity of cartographical information accumulated on the map and above all of that conditioning the construction of hydrogeographical units is possible only owing to the logical structure of the graphical elements.

The map of physico-geographical regions is a complex synthetical map representing not only the resulting constructed phenomena but even the conditioning factors, i.e. geomorphological units (relief forms are represented by colour, genesis by positive hachure), climatic regions (represented by negative hachure) and vegetation tiers (represented by dot symbols). All data serving the construction of the various physico-geographical units are graphically marked and can be found on the map similarly as among the numerous symbols attached. The repressed print of the reduced topography serves the location of the contents of the map.

The correct content, the exact construction, the harmony of means of representation and the expressional ability make the map a successful cartographical aim which can be fully utilized in planning practice and is a foreshadowing of new methods of representation and modes of compilation in thematical cartography.

Notes

In Volume 11, Chapter III in the list of authors to the article "Regional Geographic Forecast of Man's Influence as a base of Optimizing the "Man – Environment System" should be included Nevyazhsky O.I., Tatarintsev B.V. and Tikhotsky K.G.

Blažek M.
(ČSSR, Brno)

ECONOMIC REGIONALIZATION OF THE CZECH SOCIALIST REPUBLIC

The economic (economo-geographical) regionalization of
the Czech Socialist Republic was based on the conception that
the actually existing regions are always of nodal type. The
cores of the regions are concentrations of production and urban settlement. By relations of towns and their spheres of influence the movements of population (migrations, commuting of
all kinds, recently even centripetal movements, recreation
trips) are expressed.

Czechoslovakia manifests itself as the integral region
(a macroregion), a production-consumption complex of international comparison. Its core is the integral one million agglomeration of the country and its largest industrial centre.
The macroregion is divided into two national entities. Both
the republics, the Czech Socialist Republic and the Slovak
Socialist Republic, are further divided into mesoregions,
subregions, microregions. The building stones of the regionalization are the communities.

This paper does not wish to characterize the research
results but only to stress some facts of wider validity.

(1) The regionalization was carried out by a uniform
system of indices, differentiated in their meaning in harmony
with the hierarchical arrangement of the individual levels
(e.g. commuting to universities + long-distance commuting to
work, daily commuting to work + services of super-local sig-

nificance, attendance of agricultural centres + medical
centres, etc.). The cores too had to form a uniformly defined
system (million metropolis, urbanized region of metropolitan
type, industrial town of super-local significance, community
of urban or partly urban type).

(a) The spatial structure of the regionalized territory
is very heterogeneous. So, for instance, in the level of
mesoregions the influence of the cores (nuclei) did not pene-
trate in the marginal zones everywhere. Some "deaf" spaces remained. The regionalization must necessarily divide the space
investigated completely but not always in all hierarchical
levels.

(b) This statement stresses the significance for instance
for perspective planning. It defines the zones where other
centres with respective relations should be built or where
the relations of the existing centres should be strengthened
so as to comprise the areas "deaf" so far.

(2) The regionalization has shown that there exists - at
least on European scale - a common lay-out in the hierarchy
of the centres modified naturally by concrete conditions. This
allows international comparisons.

(3) The utilization of production and non-production indices apparently not corresponding is possible in two conditions:

(a) The indices are ranged in their mutual internal not
only formal succession so as to correspond in significance to
the functions of the respective regional nuclei. The methods
are generally known to consist of two stages: the initial selection of nuclei and the delimitation of the region. Both
stages are closely intercommercial.

(b) The application of the indices of various economic
branches and functions complicates the delimitation and partly
even the definition of the cores. If the region is understood
as a spatially structured unit divided into spheres closely
and freely linked to the core the periphery of the region has
the appearance of a more or less broad border zone. Inside the
zone the resulting division is to be found.

169

(c) The method of the overlaps and the resulting border zones allow an interconnection of the regions of various hierarchical levels on the basis of composition. Here the delimitation is understood when the appurtenances of the region of a lower level to a superior region is given by the levels of this region so that the links to the core of a region of lower order operate as factors leading to greater precision and linear delimitation.

(4) The steadily increasing relation between economic regionalization and administrative regionalization in harmony with the increase of significance of administrative functions for the development of economy was proved again, as well as the thesis on the uniformity of both regionalizations. This uniformity is to be seen as a process in the dialectics of their mutual relations solved with optimum results also on the basis of composition.

(5) The system approach in geography, the landscape and environment problems stress again the call for finding a correspondence between economic regionalization and physicogeographical regionalization. Our experience does not allow drawing a conclusion. The heterogeneity of the relief in a relatively uniform character of the climate caused that the defined economic regions (mainly of higher levels) correspond to a considerable extent to the physico-geographical regions in which lowlands and basins, economic cores of regions with marginal higher situated more woodded regions with prevalence of agriculture and recreation functions are combined. Even here the role of the mutual composition comes into play. As a basis of the geographical (complex) regionalization can be accepted the economic regions which are a complex of physico-geographical regions.

JORDANOV T.

(Bulgarie - Sofia)

CRITERES PERMETTANT DE DETERMINER
LA ZONALITE DE L'AGRICULTURE EN
VUE DE SA PLANIFICATION

Lors des recherches dans le domaine de la Géographie agricole une place centrale est occupée par les problèmes se rapportant à la découverte, à la connaissance et à la mise en pratique des lois qui régissent sa répartition géographique, sa structure et sa différenciation territoriale. Ces lois sont très complexes et représentent un système particulier, comprenant un grand nombre de facteurs, liés à la production, au transport et à la consommation des produits agricoles.

De pair avec l'utilisation des réalisations de la technique et de la Biologie agricoles, l'organisation territoriale (géographique) a une grande importance pour l'accroissement de la production agricole, la baisse de son prix de revient et l'augmentation de son effectivité économique. L'organisation territoriale (géographique) de l'agriculture est l'objet de la Géographie économique, de l'Economie de l'agriculture et de la Planification territoriale. Ces trois sciences se complètent mutuellement et sont en état, agissant de concert, de résoudre judicieusement les problèmes de l'organisation territoriale de l'agriculture, la délimitation de ses zones, régions, sous-régions et micro-régions.

La planification de l'agriculture, conformément aux unités territoriales citées plus haut, crée la possibilité de mettre en pratique des dispositions d'ordre social, grâce auxquelles, on peut utiliser de façon rationnelle, économiquement plus effective, les conditions et les ressources naturelles, les réalisations de la technique et de l'Agrobiologie agricoles, les ressources en main d'oeuvre, etc. Tout cela contribue à l'accroissement de la productivité sociale du travail dans le domaine de l'agriculture.

Les critères servant à délimiter les régions, sous-régions et micro-régions agricoles, sont relativement bien élucidés, mais ceux qui se rapportent à la découverte et à l'étude des zones agricoles, continuent à être insuffisamment traités par la science Ils représentent pourtant une tâche typique de la Géographie agricole et un facteur important de ses rapports plus étroits avec la Planification territoriale.

Pour découvrir la zonalité dans la répartition géographique de l'agriculture, il est indispensable de découvrir une certaine zonalité dans la répartition des conditions et des facteurs qui influent directement ou indirectement sur son développement et sa différenciation territoriale. On rapporte au groupe des facteurs qui influent directement sur la localisation, la structure et la spécialisation de l'agriculture: les conditions et les ressources naturelles, la quantité et la qualification des ressources en main d'oeuvre, le degré de mécanisation, l'emploi d'engrais minéraux et d'ingrédients chimiques, l'irrigation artificielle, l'utilisation des réalisations des sciences agrobiologiques, etc. Les facteurs, ayant une influence indirecte sur la zonalité de l'agriculture, sont en rapports étroits avec le placement de la production agricole - les centres de consommation et surtout les grandes villes, la capacité des entreprises de l'industrie alimentaire, ainsi que celle de certaines entreprises de l'industrie légère, la cofiguration du réseau des transports, le placement des produits agricoles sur les marches intérieurs et extérieurs, etc.

Plus le degré de corrélation entre la zonalité dans la répartition des conditions et des facteurs, influent directement ou indirectement sur la différenciation territoriale de l'agriculture, est grand, plus son caractère zonal peut être délimité de façon judicieuse.

L'utilisation d'une subdivision pratique par régions physico-géographiques est d'une grande importance. Dans cette subdivision on doit découvrir le potentiel naturel des zones, sous-zones et micro-zones physico-géographiques, en vue du développement de l'agriculture. Par exemple, on découvre dans la plaine de la Thrace Septentrionale et les terres avoisinentes la zonalité suivante dans les conditions et ressources

naturelles: versant nord des Rhodopes, pied des Rhodopes, vallée de la Matitza, pied de la Sredna-Gora, chaîne principale
de la Sredna-Gora. La disposition de ces unités naturelles présente un caractère zonal à potentiel local spécifique pour le
développement de l'agriculture. Voilà pourquoi, dans chacune
de ces zones il est indispensable de mettre en pratique et
d'organiser de telles mesures agrotechniques et une telle
structure de l'agriculture, qui permettent d'utiliser de la façon la plus effective le potentiel naturel respectif.

Ces bandes de terre, étroites mais longues, sont dispo-
sées de l'ouest à l'est, de la Maritza, de laquelle se trouvent les villes de Pazardjik, de Novi-Kritchim, de Plondiv, de
Parvomai, de Dimitrovgrade, etc. Ces villes qui possèdent une
industrie alimentaire très développée, sont de grands consommateurs de produits agricoles. L'industrie des conserves alimentaires y traite 40 à 60% des légumes et des fruits, l'industrie
du vin - 40 à 45% du raisin, l'industrie laitière - 82% du
lait, produits dans la plaine de la Thrace Septentrionale et
les terres avoisinantes. En outre, la population des villes
situées sur la Maritza consomme à l'état frais 75% du lait,
20% des cerises, 8 à 10% du raisin, des pommes, des fruits,
produits dans la plaine et les terres avoisinantes.

La coïncidence de cette grande zone de consommation de
produits agricoles le long de la Maritza, avec la zonalité
des conditions et ressources naturelles, a contribué ici à la
formation et au développement d'une zonalité typique dans
l'agriculture. Il s'est formé ici une zone d'agriculture intensive, avec une spécialisation dans la production de légumes, de fruits, de raisin, de tabac, ainsi que de conserves
alimentaires, de vin, de cigarettes, etc.

En vue de la découverte judicieuse de la zonalité dans
l'agriculture, ont été utilisées les données, des fermes dist-
tinctes, revatives à la structure de la terre arable, la
structure de la production globale et nette, le rendement moy-
en par décare, le prix de revient d'une tonne de produits, le
revenu net par décare, la rentabilité de la production des certaines cultures agricoles et d'animaux domestiques, etc. Le

groupement de ces données a créé la possibilité de tracer
les contours principaux de sous-zones agricoles suivantes:

1. Sous-zone typique de culture de plantes, située sur
le versant nord des Rhodopes, avec une spécialisation dans la
production de raisin, de tabac et de fruits.
2. Sous-zone typique de culture de plantes située dans
la vallée alluviale de la Maritza, avec une spécialisation
dans la production de légumes, de primeurs, de pommes, de riz.
3. Sous-zone de culture de plantes et d'élevage située
dans les parties septentrionales de la plaine de Pazardjik et
de Plovdiv, ainsi que dans celle de Stara-Zagora, avec une
spécialisation dans la production de céréales, de betterave,
sucrière, de tournesol, de coton, de viande et de lait.
4. Sous-zone de culture de plantes et d'élevage située
au pied sud de la Sredna-Gora, spécialisée dans la production
de raisin, de fruits, de viande et de lait.

Chacune de ces sous-zones agricoles se caractérise par
un ensemble spécifique de conditions climatiques et écologiques de production; par la structure, la spécialisation, les
rendements moyens, le prix de revient, le revenu net, la rentabilité de la production, les perspectives de développement
de l'agriculture, ainsi qu'un grand nombre d'indices économiques.

Tous ces indices doivent être en vue pour les organes de
la planification territoriale, afin de réaliser des mesures
concrètes, économiquement les plus efficaces, visant l'organisation territoriale de la gestion de l'agriculture dans les
zones, sous-zones et micro-zones respectives. Cela contribue-
ra à obtenir plus de production agricole, ayant un prix de
revient plus bas. De la sorte, la Géographie économique se
transforme en force productive.

Mareš J.
(CSSR, Brno)

REGIONALIZATION OF INDUSTRY IN THE CZECH SOCIALIST REPUBLIC

In the work dealing with regionalization of industry of the Czech Socialist Republic various methods were tested. The modes of regionalization carried out so far, such as grouping of industrial centres according to cooperation relations in production, grouping of districts according to statistical characteristics of similar character grouping of communities according to share of population employed in industry, etc., proved not to be suitable for detailed regionalization. A new method was applied described in English in the book "Regional studies, Methods and Analyses", Budapest 1975 and in Russian in the Volume of Papers "Regionalnaya nauka o razmeshchenii proizvoditelnykh sil", Novosibirsk, 1973.

The regionalization of industry of the Czech Socialist Republic was carried out by means of the method mentioned. From the viewpoint of applied delimitation criteria it has a rather socio-economic character. Technological and cooperation relations manifest themselves only secondarily.

On the territory of the Czech Socialist Republic altogether 1766 places with factory industry were established. A very important result was the statement that these places do not exist isolatedly but that they are connected by inter-centre commuting into territorial agglomerations of higher order, the so-called industrial nodes. All such nodes were established comprising 711 most important industrial places. Only smaller industrial centres situated in peripheral areas remained outside the nodes. It was proved that in areas where industry supplies a sufficient choice of various job opportunities for people with various qualifications the inter-centre movement and the interrelations in the nodes are increasing. The interrelations are very different and this is why no simple dependence between the intensity of relations and the distance between industrial places can be valid. The intensity of the relations cannot be established on the basis of simple distance in the Czech Socialist Republic. In grouping the industrial places into higher territorial units all concrete intercentre relations should be investigated.

It was further established that in the nodes agglomeration savings can be achieved owing to higher exploitation of labour, savings in housing construction and public equipment, savings from a better utilization of the existing infrastructure, etc. A purposeful utilization of the nodes is accordingly one of the ways of increasing the social efficiency of expenditures.

The work has proved that the gravity territories around the industrial places are very heterogeneous. They have a various extent, intensity of gravity and compactness. The degree of effects of industry is also different in the various gravity territories. But they are a relatively stabilized territories changing their extent not even case of longlasting changes of the rate of employment in the core. Their character is affected by several agents above all by the geographical position of the industrial core, the population density, the settlement structure, the communication network, etc. It was stated that there is no generally valid scheme of the relation between the size of the centre and the size of its hinterland mentioned in several theoretical works.

By addition of the gravity territories to their cores which were in the simpler case the various industrial places, but usually the industrial nodes, altogether 155 industrial territorial units were defined. According to their characteristics the units were divided into two basic types (industrial areas proper and peripheral industrial territories), subdivided into 5 subtypes. Thus 11 principal, 23 more important and 12 other industrial regions and 21 larger and 88 other peripheral territories came into origin.

The method applied resulting in the delimitation of the types of industrial regions in the Czech Socialist Republic has simplified substantially the image of the distribution of Czech industry. It contributed to the objectivisation of

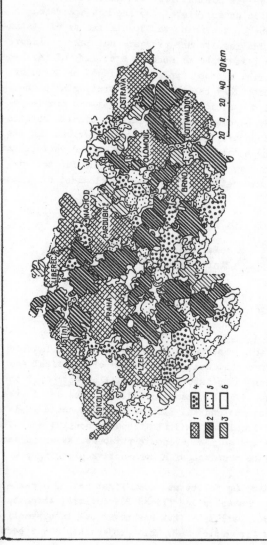

Industrial regions of the Czech Socialist Republic:
1 – principal; 2 – more significant; 3 – other
peripheral industrial territories; 4 – larger; 5 –
other; 6 – territories outside the industrial regions

the defined territorial units grouped formerly not according
to industrial places and communities, but according to much
larger administrative units (districts) and not according to
the mutual relations but according to the similarity of statistical quantities. The results achieved are not only of theoretical significance (existence of nodes), stability of the hinterlands of industrial places, types of regions) but even of
practical importance. They can be used in territorial planning, geographical and sociological investigations and in pe-
dagogy. They should become a base for further studies concerning the achieved economic level of industrial regions and the
main disproportions between local resources of the regions and
the demands of industry for these resources, and for the elaboration of a recommendation for further development of industrial regions.

Merlini G.
(Italy, Bologna)

GEOGRAPHICAL IMPLICATIONS OF NATIONAL AND
REGIONAL PLANS IN ITALY

1. The _first stages_ in the Italian planning policy
("Piano Vanoni"; acts concerning freeways and ports; laws for
the Mezzogiorno) are just made of decisions of political and
social nature which carry into effect a _will of action_ of the
government and the official economic bodies. These act therefore on infrastructures and only incidentally on human landscape, except for irrelevant modifications or adaptations of
the natural landscape to meet economic purposes. Nevertheless,
these stages are the premises of a more active planning policy.

2. A real planning policy has been first set on a political ground with the so called "Piano Pieraccini", which has
become a law valid until 1969, but was never put into practice
and also with the "Progetto '80". This project has never been

expressed in form of an act, but it is accompanied by a detailed documentation including cartographic material and in fact has been so far at the root of the planning policy of the special governmental Committee for Economic Planning and other official bodies as the Secretariat for Special Interventions in the Mezzogiorno, Cassa del Mezzogiorno, Regions, Provinces, and Communes which all have regulating powers for the formulation and fulfilment of development plans.

The two mentioned plans however are not binding for private firms but only largely orientating. They mainly rely on incentivation which, on the other hand, was also stressed in all the acts passed in those same years and concerning infrastructures, financial instruments and competences of local governments. On the contrary, the Plans are binding for State firms.

The two plans have effected the Italian landscape with infrastructures such as new roads, freeways, airports and ports, with reclaiming and irrigations, power plants and, as far as plants' location is concerned, with the so called policy of poles and industrial areas and with the agrarian policy of the unions for reclamation and of some local bodies. In these conditions, the plan's action only effects single dots or limited areas and never extends over regional areas, nor it effects the way of life. The transformations of landscape are therefore limited to a spatially-limited and univocal pattern.

3. The establishment of the "Regioni", with legislative power specifically referred to plans for socio-economic development (plan's regionalization), although it has maintained national bodies and the formulation of a national plan within which the "Regioni" are to accomplish their own plans, has brought to more functional and geographic outlook of the plans. In fact, a few general lines and instruments of sectorial and territorial development are still under State competence, while every "Regione" is responsible of the establishment or restoration of a well balanced development within its territorial limits, and the defining of its own functions within the national community.

The planning action, still in the difficult situation of a crisis that may be estimated among the hardest and long-lasting in the recent Italian history, shows in its actual limits a picture of both functionality and geographicity.

Regional plans have a _functional_ outlook in the way that actual features of the landscape of single regions are realized and by means of proper interventions (environmental studies or special projects as the singling of axis of development and the study of the irradiating power of existing poles and nodal points through established infrastructures) are given a _development_ in terms of _further anthropization_.

Truly, plans have also a _geographical_ outlook in the way they assimilate geographical concepts as those of _region_ (although the "Regioni" not always have limits overlapping with those of geographical regions), of landscape (although the physical features are emphasized, thus stressing conservation of natural environments and neglecting rational forms of utilization of the landscape as a whole, in relation to its anthropic features) and the concept of _integrated development_ in terms of promotion of better conditions of life in all regions.

4. A few of regional plans (Emilia, Lombardia, Trentino-Alto Adige and the Mezzogiorno in general) explain the functional and geographical outlooks of the plans in their actual formulation, also showing that a further anthropization of the landscape is long more difficult than colonization _ab imis_ of virgin lands, which is a radical transformation of the landscape, especially if action is left to private enterprises, instead of submitting them completely to the plan policy or to overwhelming power of great monopolies and multinational enterprises.

Shafi M.
(India, Aligarh)

ASSESSMENT OF VON THUNEN'S LAND USE ANALYSIS IN
INDIA

A major shift in the technique of land use analysis has
taken place in the late sixties where more specific attention
has been given to problems of land values and the intensity
of farming in relation to varying distance from the settlements. Empirical studies have shown complex relations between
the simple distance base models and intricate factors like
cumulative causation and cultural barriers. Attention is now
given to the determination of the best combination of in-puts
in a given level or several levels of production or evolving
the best combination of land uses with fixed in puts levels.
Instead of comparing a series of maps in the traditional
style, regression equations have been developed to express
relations between a number of spatially distributed variables.
The technique of correlation regression analysis is now being
widely applied in land use studies to show how closely any
two variables are related by a simple correlation and how
much of a change in any variable is associated with the corresponding change in another variable.

In this paper the author has built up a land use model
to examine the influence of different variables including
distance from the settlements on the location of land use
pattern and has arrived at the conclusion that, in a country
where farming is the main stay of the people, it is chiefly
the availability of irrigation facilities and not distance
which influences the location of the pattern of crops.

A comprehensive land use study of about 35 villages in
Koil Tehsil, District Aligarh, (India) was carried out. Attention was focused on percentage of different crops in each
season, availability and intensity of irrigation facilities,
and intensity of double cropping. These data were obtained
from the village studies conducted by the students of the De-

partment of Geography, Aligarh Muslim University as Project
Work during their postgraduate studies.

In order to see the impact of distance from urban settlement on intensity of land use and cropping pattern, the impact
of irrigation facilities and intensity of irrigation on double
cropping, and the types of crops to be grown, the following
types of zoning pattern were considered:

(i) Zoning around the urban settlement of Koil Tehsil by
taking Railway Station as the Centre. In this case decision
was taken to draw concentric zones at an interval of one mile
around the periphery of urban land. Study was conducted upto
a distance of nine miles (14.4 km) and hence nine circular
zones were drawn. In each zone data on the percentage of cultivated land for different groups were collected, including
Rabi and Kharif seasons. The crops were grouped as food
grains, vegetable crops and cash crops.

(ii) Zoning around the Canal:

In this case eliptical zones were drawn at an interval
of 0.4 km (1/4 mile) by taking central line of Ganga canal as
the major axis. Only seven zones were considered for study
purpose. This study was related to determine the impact of
intensity of irrigation on intensity of land use and cropping
pattern.

(iii) Zoning on the Basis of Number of Tubewells:

From the irrigation map supplied by Minor Irrigation Department it was found that in the South-West of Tehsil Koil,
the farmers had to rely upon tubewell irrigation facilities
as there is no canal irrigation facility available there. A
network of State Tubewells can easily be found with a radius
of influence on one mile (1.4 km) approximately. In order to
analyse the effect of tubewell irrigation facilities on cropping pattern and double cropping system, conical zones with
centre as Railway Station of Tehsil Koil were considered.
Radial lines upto periphery of five miles (8 km) from the
centres were drawn at equi-angle so that equal area is en-
closed in each conical zone.

After evaluating the percentage of cultivated land utilized for each crop group in the corresponding zone, in the case of each zoning pattern (Rabi as well as Kharif), the curves are plotted with the ordinate representing the percentage of crop group and abscissa as the distance from urban settlement in the first type of zoning, distance from the major axis of canal in the second type of zoning, and number of tubewells in each zone in the third type of zoning.

The following conclusions were drawn from the graphs for different zoning patterns:

(a) Circular Zoning Around the Urban Settlement:

For all types of crop groupings (food grains, cash crops as well as vegetable crops), the percentage of cultivated land utilized with distance from the settlement does not hold any relationship. The curve trends are in zig-zag form with increasing and decreasing order. For example in case of food grains in Kharif season, the percentage of land utilized in the first zone is 68%, in the second 57%, while in the third zone it has increased to 85%. In the last zone which is very far from urban area it is found to be 63% only. Similarly in case of cash crop in Kharif season, the percentage of land utilized is decreasing with the distance from urban area upto fifth zone and then it has increased tremendously.

Study related to double cropping also does not follow Thunen's Principle strictly because according to Thunen's Principles the intensity of land use should decrease as the distance from the urban area increases, while in this case the intensity of land use with the distance from the settlement does not have any relationship.

(b) Eliptical zoning around the canal:

In this case it is found that the percentage of the land utilization for different crop groups except food grain in Kharif is decreasing as the distance from the canal is increasing, for example the percentage of cash crops, which near the Ganga canal is 25% of the total utilization and it is 23, 15 and 13% in the second, third and fourth zone respectively or we can say that it decreases with increasing distance from the Canal.

Double cropping pattern has a decreasing trend as the distance from the canal increases. Mathematically we can conclude that the percentage of the land utilization in double cropping pattern is invariably proportional to the distance from the availability of water resources used for irrigation. Hence   Thunen Principle can be modified as follows:

"The intensity of land uses decreases with
    increasing distance from the irrigation
    source rather than from urban settlement".

(c) Zoning on the basis of the number of tube-wells:

In this case it is assumed that the irrigation capacity of the tube-wells is the same. It is seen in this zone that the intensity of land utilization for vegetable and cash crop decreases with decreasing number of tubewells in the zones of equal areas. Similarly double cropping pattern has a decreasing trend as the availability of the amount of ground water decreases. Kharif crops are not effected by decreasing number of tubewells in the respective zones because the food grains in the Kharif seasons are mainly a function of the type of land and intensity of rain falls within the region.

A comprehensive study was carried out for specific villages such as Begpur, Lekhrajpur, Salimpur Maafi and Zorawar of Tehsil Koil in the following systematic ways:

(a) The settlements of the villages were considered as the centres and with respect to the centres, the lands surrounding the villages were zoned.

(b) The total land under each zone was computed by using graph papers and counting the squares on graph paper on each zone.

For example, the total land covered in case of village Begpur from zone 1 to 6 are 12, 42, 80, 24, 24 and 25 units respectively.

(c) The area used for various rabi crops such as foodgrains, cash crops, vegetable, fodder and double cropping were computed for each zone and the percentage of the total

184

area for each crop group was calculated. For example, in case of village Begpur, the area utilized for foodgrains was 5 units which is 41% of the total area of 12 units.

(d) In order to see the impact of irrigation and double cropping pattern on the distance of urban settlement, the area under double cropping and the total area to be irrigated were computed by using various charts prepared under the land-use survey for specific villages.

The following conclusions were drawn:

(1) With respect to the distance from the village settlement the intensity of land used for various rabi crops such as foodgrains, vegetables and fodders, does not have any relationship. For foodgrains, the maximum intensity of land utilized is found in zone V while minimum is in zone IV. Similarly for vegetables, maximum intensity of land used is 8% in zone III while minimum as 1% in zone II. In the same manner for fodders, the minimum intensity is 1% in zone II while maximum as 4% in zone I.

In case of specific villages Lekhrajpur, Salimpur and Zorawar with respect to the distance from the settlement of respective villages, the intensity of land utilized for wheat, wheat-gram, peas and double cropping does not show any direct or indirect relationship but with increase in irrigation facilities vegetable and double cropping increase in intensity.

The study thus clearly shows that the intensity of land use in the villages of Koil Tehsil does not show any relationship with varying distance from the settlements - instead it shows a relationship with varying distance from irrigation facilities.

Woldenberg M.J.
(USA, Buffalo)

A PERIODIC TABLE OF SPATIAL HIERARCHIES

A computer algorithm was developed to examine all possible mixed hexagonal hierarchies of up to ten orders. Five tically found target hierarchies were used to test alternative models. No model which generates a "best" geometric progression of areas (no matter how "best" was defined) was successful in hitting the target hierarchies. A linear model, which creates the best geometric progression of lengths (proportional to the square roots of hexagonal areas) was successful in hitting the targets.

Tables of eleven hierarchies are presented here. Five of these are the empirically found target hierarchies. Two more are hierarchies which may have been recognized before. Four new hierarchies are proposed, which have not yet been recognized. New empirical data for one previously unpublished hierarchial type is presented here.

VII. GEOGRAPHY OF POPULATION

Russel B. Adams
(USA, Minneapolis, Minnesota)

THE SOVIET METROPOLITAN HIERARCHY; REGIONALIZATION AND COMPARISON WITH THE UNITED STATES

The rapid growth of Soviet cities is converging toward a hierarchy similar to that of the United States. The number and aggregate populations of metropolitan centers by five size categories in the two countries are compared for growth and change from 1939 to 1973. Also, nine Soviet urban regions are identified, mapped, and correlated with comparable American groupings. Growth rates of Soviet metropoli are normalizing with less recent variation as compared to the 1939-1959 period, a trend which parallels that in the United States. Also, it appears that certain functions, such as administration and transportation, are stabilizing factors in urban growth.

Governmental policies of investment in underdeveloped regions, "balanced growth", and diversification may be partially thwarted by five-year planning goals which have stimulated supra-growth in large cities of the South and East. However, it seems likely that increasing mobility, amenities, and the expansion of consumer goods and services will produce a reversal of trends toward higher growth rates in the metropolitan centers of the West. Projections to the Year 2000 suggest that Soviet metropoli will have a larger share of the national population and a more uniform growth pattern than those in the United States.

Alexander, C.S. and Monk, J.J.
(USA, Urbana, Illinois)

THE CHANGING RURAL LANDSCAPE IN WESTERN PUERTO RICO
1950 - 1975

Changes in the traditional rural scene, as the result of modernization, have been virtually ignored, not only in Puerto Rico but in all Latin America as well (D.A.Preston, 1974; J.J.Parsons, 1973; P.A.D.Stouse, 1971). In Puerto Rico modernization is resulting in abandonment of agricultural land, the emergence of new migration paths, in addition to the standard rural-urban mode, and a new pattern of rural settlement. The island is ideal for a study of these kinds of change because detailed geographic and anthropologic field studies made in the late 1940's provide an accurate base line of information that makes such a task possible (C.F.Jones, 1950-1951; C.F.Jones and R.Pico eds., 1955; J.H.Steward, 1956; H.R.Imus, 1951).

Dramatic changes have been taking place in rural Puerto Rico during the last twenty five years. One obvious change has been from a dominantly rural population to one that has become predominantly urban. In addition there has been a substantial redistribution of population on the island. The north and east, an area of urban-industrial and agricultural growth has experienced a marked increase. In contrast, the western hill region, an area of declining coffee production, has been losing population since before 1910, and the decline has continued.

Though the area is one of overall population decline, when it is examined closely the pattern of population change is complex and the processes of change are more varied than one of rural-urban migration. Of the thirty-nine hill <u>barrios</u> (wards) around Mayaguez, ten gained population constantly, thirteen lost steadily and sixteen experienced fluctuations. Of the latter, eleven gained and five lost overall (U.S.

Bureau of Census, 1960-1970). There is a strong association between barrios of declining population and dependence on agriculture for livelihood. In addition, residents of these barrios relatively rarely go to town. Those barrios with the greatest and most consistant growth have virtually no one working in agriculture, the employment rate is high, a majority of those employed are working in town and the residents regularly go to town for shopping. The gaining barrios also have direct, well established roads to Mayaguez whereas the declining barrios are more distant. The nature of land ownership is another factor influencing population change. Population decline occurs in those barrios where large landholders have ceased their operations but hold the land. Where small holdings existed or where the landholders are willing to subdivide, a potential for population growth exists.

In 1950 the principal economic activities in the western hill area were the cultivation of coffee and sugar cane and subsistence farming. These and the physical setting were conducive to a dispersed settlement pattern. In the eastern part, coffee was the predominant crop. On the larger holdings, or haciendas, the workers lived in small clusters of dwellings near the house of the owner or manager. Smaller fincas were dispersed and their employees lived in dwellings on scattered plots provided them by the owners. Towards the western side of the region, sugar cane cultivation extended from the lowlands well into the hills, even occurring on steep slopes. The larger cane fields did result in some settlement concentration, though again in small clusters of houses rather than agricultural villages. West of the present Highway 2, subsistence farming predominated on steep, eroded land. The population here was dispersed on ridge tops, slopes and in narrow valleys.

Most of the rural population in 1950 had little opportunity or reason for travel. Rugged terrain and poor roads in both the coffee and subsistence cultivation areas restricted movement. In the cane area the roads were probably better but focused on the mill. The country people did make a few

189

trips to town during the year for clothing, hardware, some dry goods and various celebrations. Food was locally grown or purchased from hacienda stores. Rural stores were rare, and those that did exist emphasized liquor sales and recreation.

During the last twenty-five years, the government of Puerto Rico has sponsored many industries throughout the island as well as having provided rural water supplies, electrification and many new, paved roads. These modern improvements are changing the rural landscape in two significant ways. One is a change in land use, the other is increase in the number of houses along paved roads. Land use changes result from abandonment of sugar cane and a severe decline in coffee production. The cane fields of 1950 are now unused areas of poor grass and brush, secondary forest, or have been planted to grass for dairy cattle grazing. Change among the coffee plantings is more complex. Some are still producing and are reasonably well kept. Others are minimally tended and are poor in productivity. These often grade imperceptibly to coffee groves that are completely abandoned to forest.

The reasons for the land use changes are fundamentally the same - low wages and seasonal unemployment set against a period of increasing industrialization. Regular and higher paying employment resulting from the island government's policy of attracting industry to Puerto Rico made it increasingly difficult for planters to obtain the labor they needed. As a result many coffee growers and virtually all the cane growers in the uplands simply ceased to operate.

Change is also taking place in the rural settlement pattern; an increase in house construction along the new paved roads. As a consequence there is emerging a linear housing arrangement in barrios of decreasing as well as those of increasing population. This new pattern is very different from the dispersed dwelling distribution characteristic of the area prior to 1950 (C.F.Jones, 1950-1951). The most frequent reason given for this move is to improve residential quality (J.Monk and C.S.Alexander, 1975). Governmental projects encourage new housing by providing money for construction mate-

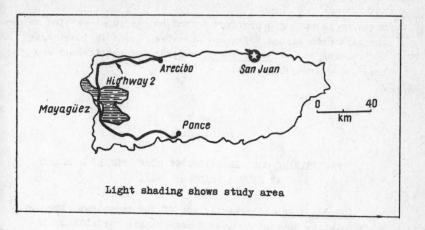

Light shading shows study area

rials through low interest loans if the home owners will provide mutual construction assistance (R.Pico, 1975). Changing
family status and moving to be near relatives are also reasons
for new building. Here family assistance in construction is
likely. Other reasons for moving are changes in employment,
desire for improved facilities and, a preference for the rural
environment. The latter attracts residents disenchanted with
the urban setting because of cost, or perceived pollution and
crime. Many of these new rural settlers are industrial workers from Mayaguez, migrants returning from the United States
or retired small business men.

Several interesting facts are associated with increasing
roadside settlement. Sample survey data show about 57% of the
roadside residents made only short within barrio moves. However, a large number (ca. 43%) of the people moving to the
roads come from outside barrios or urban areas in western
Puerto Rico. This rural to rural and urban to rural movement
does not fit the commonly held theme of rural to urban migration in Latin America.

The combination of land abandonment associated with increasing industrialization and, a steady movement of people
to new roads because of the accompanying advantages of moder-

nization is having a profound effect on the rural setting in
the hill area around Mayaguez. The rural scene is increasingly becoming one of strassendorf residential settlements and
inter-road abandoned land.

Ayyar N.P.
(India, Sagar)

PRELIMINARY CONSIDERATIONS ON RURAL FEMALE LITERACY IN MADHYA PRADESH, INDIA

Introduction: Literacy is one of the important elements
of population composition. As a demographic variable it indicates the level of socio-economic development of the community concerned and influences and is influenced by other demographic variables like occupation, caste and community composition, age- and sex-composition, etc. Although one could
study both male and female literacy and both urban and rural
literacy, in a country like India rural female literacy has
special interest. The lack of education has been both a cause
and effect of rural backwardness; while with a dominance of
males in literacy and the increasing percentage of female literacy (from 3.36 in 1961 to 6.10 in 1971), the study of rural
female literacy is sure to prove of great interest to demogra-
phers and population geographers.

General Facts: For Madhya Pradesh as a whole, the literacy percentage amongst the rural female population was 6.10
in 1971, this being almost the lowest among the States and
Union Territories of the country. Female (rural) literacy in
the State as elsewhere in the country, is less than male literacy (27.0); also it is less than female (urban) literacy
(37.0).

Amongst the districts of the State, West Nimar had the
median figure of 5.11 (per cent of literate to total rural
female population); Panna was at the lower quartile with 3.84%
and Damoh at the upper quartile with 7.89 per cent.

The percentages for the 190 tahsils of the State vary
from 0.9 for Konta tahsil (Bastar district) to 13.9 for Narsimhapur tahsil (Narsimhapur district).

Distribution of Rural Female Literacy: The 190 tahsils
of the State could be divided into 7 categories as follows:

S.No.	Category	Percentage of Literate to Total Rural Female Population	Remarks
1	Very Low	Below 1.60	x - 2 S.D.
2	Low	1.60 - 2.91	x - 1½ S.D.
3	Lower Medium	2.91 - 4.23	x - 1 S.D.
4	MEDIUM	4.23 - 6.86	x ±½ S.D.
5	Higher Medium	6.86 - 8.18	x + 1 S.D.
6	High	8.18 - 9.50	x + 1½ S.D.
7	Very High	Above 9.50	x + 2 S.D.

(x - Mean of 190 figures
S.D. - Standard Deviation)

The map of Rural Female Literacy percentages shows three
prominent belts of High values: (I) Along the Narmada Valley,
(II) Parts of Satpuras in south Madhya Pradesh, (III) Parts
of Chhattisgarh and (IV) a lone part in northeast Madhya Prade
Pradesh. Low values are found in the tribal districts of Bastar, Jhabua, Shahdol, etc. Values are generally low northwest
of the Narmada-Son Axis and also in northeast Madhya Pradesh.

Causes for the prevailing rate of literacy in Madhya
Pradesh: The causes for illiteracy are generally to be sought
in the socio-economic and cultural background of the region.
Among the various factors may be mentioned: the occupational
structure and standards of living, the community-wise composition of the population and the availability of school facilities. In the case of female education, prevalent prejudices
are an important factor. This factor varies not only in space,
but also with time. The fact that education has held very
little functional value for the traditional subsistence agricultural economy of the region in question is also to be kept
in mind. Early marriages among girls, the general neglect of

females in the rural areas and the prevalence of the 'pardah' system in certain families are also important factors.

Causes for the Regional Variations in Female Literacy Rates: The map of Rural Female Literacy percentages may be compared with maps of (I) Rural Scheduled Caste/Tribe population, (II) Rural Sex-Ratio, (III) Rural Agricultural Population and (IV) Rural Male Literacy.

It can generally be seen that areas with a HIGH percentage of Scheduled Caste/Scheduled Tribe (S.C./S.T.) population have a LOW female literacy (F.L.) percentage. This is because people belonging to these communities are generally quite poor and cannot afford to send their daughters to school. Some examples at district level are as follows:

District	F. L. percentage	Rank in State	S.C./S.T. percentage	Rank in State
Jhabua	2.13	Lowest	92.5	Highest
Bastar	2.94	3rd Lowest	73.6	2nd Highest
Surguja	3.14	5th Lowest	63.8	5th Highest
Balaghat	10.80	2nd Highest	18.8	3rd Lowest
Hoshangabad	9.37	3rd Highest	21.0	6th Lowest

There are, of course, exceptions to this rule (e.g. Rajgarh, mandla, etc.)

At tahsil-level, the two maps of F.L. and S.C./S.T. show resemblance in an interesting manner. Except for two tahsils, all the tahsils which have a High percentage of S.C./S.T. population figure as 'Very Low', 'Low', 'Lower Medium' or 'Medium' in the F.L. map. The reverse does not necessarily hold good, since low female literacy may be due to other factors also.

Very generally speaking, a relatively High Female Literacy is associated with a relatively High Male Literacy (M.L.); there is no tahsil with High F.L. and Low Male Literacy. Similarly, tahsils with Low F.L. generally have Low M.L. also. There are no Low F.L. tahsils with High or Very High Male Literacy. It need however hardly be stressed that 'Low' or 'High' figures for M.L. and F.L. vary considerably between each other; in terms of actual percentages, a VERY

HIGH for Female Population (9.50) corresponds to a VERY LOW for Male Population (10.00).

Other Factors: It can also generally be surmised that female literacy rates are higher where there are a larger number of people engaged in occupations other than agriculture - e.g. services, commerce, industry, etc. Such people generally lead higher standards of living and are rather more progressive in their outlook on women's education.

Further, where the number of daughters in a (rural) family is large, the tendency is to send only the first one or two of them to the school. In other words, a higher sex-ratio could mean a lower female literacy rate.

These relationships are not generally noticeable on a tahsil-level map; but detailed field work covering some 125 families of (rural) Sagar tahsil (Sagar district) have borne out the above-mentioned significant connections. Such field work has also pointed out the relationship between income and female literacy. For instance, among these 125 families, 75 per cent of the illiterate females belonged to families getting an income of less than Rs 250 per month.

Progress of Rural Female Literacy in Madhya Pradesh: Both in terms of actual numbers and percentage, female literacy has shown an upward trend in the period 1961-71. It is interesting to note that while rural population as a whole had increased only by 24.8 per cent from 1961 to 1971, rural female literate population had grown from 458,335 in 1961 to 1,040,606 in 1971, i.e. by over 217 per cent. The per cent of literate to total (rural, female) population had thus nearly doubled from 3.36 to 6.10. This increase was shared by all the districts of the State almost uniformly, so that the relative ranking of most of the districts was nearly the same in 1971 as it had been in 1961.

Bohra Dan Mal
Jodhpur (India)

PATTERNS OF URBAN SETTLEMENTS IN RAJASTHAN (INDIA)

Out of 26 districts, towns in 22 districts are approach-ingly uniform in distribution. Throughout the State of Rajas-than, urban settlements range in population from 2,655 persons up to 615,258 persons (1971 Census). In the present paper at-tempts have been made to apply the technique of "near-neigh-bour statistic" for the quantitative expression of urban set-tlements in Rajasthan. In this technique straight line dist-ance separating urban settlements has been taken into consi-deration. For details about the technique of near-neighbour statistic the reader should refer the works of Clark and Evans and of King.

According to the law of mathematical probability mean distance (rE) between each point and its nearest neighbour which could be expected in a random distribution is equal to $1/2p^{-1/2}$, where p is the observed density of points in the area under consideration. The derivation of the rE value in-volves consideration of the Poisson exponential function. The ratio of the observed mean distance (rA) to the expected va-lue (rE), is regarded as near-neighbour statistic. Table 1 shows near-neighbour statistics and nature of pattern in dif-ferent districts of Rajasthan. If the ratio between observed and expected mean distance is 1; less than 1 and more than 1, the distribution of points is said to be random, aggregated and uniform accordingly.

A precise expression of distribution pattern of towns in each district has been obtained by near-neighbour statistic (R) and test has been employed to know the significance of differences between them. Table 2 shows test for uniform spacing in different districts of Rajasthan.

Table 1

NEAR-NEIGHBOUR STATISTICS

Study area	Number of Towns	Density of Towns per square kilometer	Mean observed distance (kilometers) r_A	Expected mean distance (kilometers) in random distribution r_E	Near-Neighbour statistic, R	Nature of pattern
1. Ganganagar	12	.00058	34.17	20.75	1.65	Approaching uniform
2. Bikaner	6	.00022	16.00	33.68	0.48	Aggregated
3. Churu	11	.00065	24.55	19.59	1.25	Approaching uniform
4. Jhunjhunu	12	.00202	15.33	11.12	1.38	Approaching uniform
5. Alwar	4	.00048	34.00	22.80	1.49	Approaching uniform
6. Bharatpur	9	.00111	26.67	15.05	1.77	Approaching uniform
7. Sawai Madhopur	6	.00057	22.66	20.93	1.08	Random
8. Jaipur	11	.00079	25.09	17.78	1.41	Approaching uniform
9. Sikar	7	.00091	20.00	16.56	1.21	Approaching uniform
10. Ajmer	8	.00094	23.00	16.30	1.41	Approaching uniform
11. Tonk	6	.00083	29.00	17.35	1.67	Approaching uniform
12. Jaisalmer	2	.00005	100.00	70.66	1.42	Approaching uniform
13. Jodhpur	4	.00018	60.50	37.24	1.62	Approaching uniform
14. Nagaur	8	.00045	38.75	23.55	1.65	Approaching uniform
15. Pali	6	.00049	20.33	22.57	0.90	Random
16. Barmer	2	.00007	72.00	59.71	1.20	Approaching uniform
17. Jalor	2	.00019	56.00	36.25	1.54	Approaching uniform

Table 1, continued

Study area	Number of Towns	Density of Towns per square kilometer	Mean observed distance (kilometers) rA	Expected mean distance (kilometers) in random distribution rE	Nearest-Neighbour statistic, R	Nature of pattern
18. Sirohi	5	.00097	20.40	16.04	1.27	Approaching uniform
19. Bhilwara	4	.00038	41.00	25.63	1.60	Approaching uniform
20. Udaipur	6	.00035	37.33	26.70	1.40	Approaching uniform
21. Chittaurgarh	7	.00065	32.00	19.60	1.64	Approaching uniform
22. Dungarpur	2	.00053	42.00	21.70	1.94	Approaching uniform
23. Banswara	2	.00040	40.00	24.98	1.60	Approaching uniform
24. Bundi	4	.00072	36.00	18.62	1.93	Approaching uniform
25. Kota	6	.00048	45.00	22.81	2.00	Approaching uniform
26. Jhalawar	5	.00080	13.80	17.67	0.78	Approaching uniform

Table 2

TEST FOR UNIFORM SPACING

S. No.	Study area	Density of Towns per square kilometer	Mean observed distance rA (kilometers)	Expected mean distance in uniform distribution of given density rU	Difference between rA and rU significant
1.	Ganganagar	.00058	34.17	44.59	Significant
2.	Bikaner	.00022	16.00	72.40	Significant
3.	Churu	.00065	24.55	42.11	Significant
4.	Jhunjhunu	.00202	15.33	23.89	Significant
5.	Alwar	.00048	34.00	49.01	Significant
6.	Bharatpur	.00111	26.67	32.22	Significant
7.	Sawai Madhopur	.00057	22.66	44.96	Significant
8.	Jaipur	.00079	25.09	38.20	Significant
9.	Sikar	.00091	20.00	35.59	Significant
10.	Ajmer	.00094	23.00	35.01	Significant
11.	Tonk	.00083	29.00	37.27	Significant
12.	Jaisalmer	.00005	100.00	148.40	Significant
13.	Jodhpur	.00018	60.50	80.02	Not significant
14.	Nagaur	.00045	38.75	50.62	Not significant
15.	Pali	.00049	20.33	48.51	Significant
16.	Barmer	.00007	72.00	128.30	Significant
17.	Jalor	.00019	56.00	77.91	Significant
18.	Sirohi	.00097	20.40	34.47	Significant
19.	Bhilwara	.00038	41.00	55.09	Significant
20.	Udaipur	.00035	37.33	57.38	Not significant
21.	Chittaurgarh	.00065	32.00	42.11	Significant
22.	Dungarpur	.00053	42.00	46.63	Significant
23.	Banswara	.00040	40.00	53.68	Significant
24.	Bundi	.00072	36.00	40.01	Not significant
25.	Kota	.00048	45.00	49.01	Not significant
26.	Jhalawar	.00080	13.80	37.96	Significant

Areal differences in physical features, agricultural conditions, transport network, political factors and history of settlements influence variations in the magnitude of R values from one district to the other. It is seen in Table 1 that R value varies from 0.48 and 0.78 (showing aggregated pattern of urban settlements) in Bikaner and Bnalawar districts respectively through 0.90 and 1.08 (showing random pattern of urban settlements) in Pali and Sawai Madhopur districts respectively to 1.93 and 2.00 (showing approachingly uniform distribution of urban settlements) in Bundi and Kota districts.

The most marked tendency toward uniform spacing is noticed in Kota district. Out of 6 towns, 4 towns namely, Indergarh, Kota, Ramganj Mandi and Baran are almost equispaced. General uniformity in physical features and cultural landscape - thicker mantle of fertile soils, higher rainfall in summer months, better irrigation facilities throughout the year and a larger dispersal of population with very little uncultivated spaces - has contributed to the hexagonal arrangement of towns in Kota district. The pattern of urban settlements in Bundi district (R=1.93) reflects the importance of same geographical factors which favour even spacing of towns in Kota district.

Elsewhere in the State of Rajasthan the tendency toward even spacing is more marked in Dungarpur (R=1.94), Bharatpur (R=1.65), Ganganagar (R=1.65), Chittaurgarh (R=1.64), Jodhpur (R=1.62), Bhilwara (R=1.60), Banswara (R=1.60), Jalor (R= =1.54), Alwar (R=1.49), Jaisalmer (R=1.42), Jaipur (R=1.41), Udaipur (R=1.40) and is of less marked significance in the remaining districts particularly Barmer, Sirohi, Sikar and Churu.

There are two districts where the distribution of settlements is apparently random. For example, the values of R for Pali and Sawai Madhopur districts are 0.90 and 1.08 respectively. The towns of Bikaner district are genuinely aggregated. Out of 6 towns in Bikaner district, 5 towns show the most compact pattern.

Attempts have been made to correlate near-neighbour statistic with percentage of total population classed as agriculturists, percentage of cultivated land, density of villages per square kilometer but correlation was not found to be significant for any of the independent variables. It should be emphasised here that patterns of town distribution in Rajasthan are more related to the variations in culture patterns which had produced typical urban forms and patterns, depending both on geographical environment and social heritage than to the proportion of cultivators to total population and percentage of cultivated land to total land.

Brawer M.
(Israel, Tel Aviv)

CHANGES IN VILLAGE SITES AND PATTERNS IN MARGINAL SEMI-ARID AREAS

Over many centuries, in some cases milleniums, village sites and patterns hardly changed in the semi-arid regions of the Middle East and the Mediterranean Lands. This is particularly true of the marginal areas of these regions where the frontiers of rural settlement had been vacillating, since ancient times, under the impact of climatic cycles and desert tribes. The fact that these marginal areas had been the arena of confrontation between the sedantery peasant population and the nomadic pastoral tribes has given the villages special characteristics. These are reflected in topographic properties of their sites, in their extremely clustered pattern and their comparatively large population. Strategic advantages and requirements played here a dominant role.

The processes of transformation to which most of the rural areas, in the above mentioned regions, have been subjected in recent years, are having far reaching effects on the special characteristics of these villages. The nature of these changes has been the subject of a study, undertaken by the Department of Geography at the University of Tel Aviv. It cover-

ed marginal Arab rural settlement in Samaria and Judea. The
main processes of transformation, as brought-out by this study
can be taken as representative of developments in other similar regions of the Middle East.

Three main trends are apparant in the change of geographical characteristics which the villages surveyed, are under-
going. 1. A gradual shift of the village to a new site with
a corresponding abandonment of much of the old site. In most
cases the village, which for many generations had been confined to a hilltop, is now "creeping" down towards the valley. In some cases, over a period of 2-3 decades, the majority of the inhabitants moved to a site well below the old
village. 2. A gradual transformation from an extremely nu-
cleated to a dispersed pattern. For centuries the entire
built-up area of these villages formed a single dense cluster
of houses surrounded by small walled courts. In recent years
this is rapidly being replaced by a dispersed pattern. In
some villages this process of dispersion runs concurrently
with the shift in site, in others it is an expansion of the
old site. 3. The emergence of a growing number of hamlets or
ecarts on the lands of the village, at distances varying from
several hundred metres to 5-6 kilometres from the chief settlement. Almost the entire population of these hamlets or
ecarts came from the "mother village".

The effective imposition of law and order in the marginal areas of Samaria and Judea, over the last 4-5 decades, and
the sedentation of most of the nomads in this and in neighbouring regions have made superfluous the traditional secu-
rity requirements of the villages. The strategic advantages of
hardly accessible hilltops and densely clustered dwellings
have become serious disadvantages under the new, present day,
conditions. The villagers soon realised that much unnecessary
hardship and wasted physical efforts can be avoided by moving
their homes down into the valley, much closer to their fields
and plantations or to a modern paved road. This prompted a
growing number of villagers, especially young people, to
build new homes outside the old village on a more comfortable

and easily accessible site. There is a growing tendency to build these new home along the road leading into or passing near the village. Thus the old extreme form of "Haufendorf" is often being transformed into a dispersed "Strassendorf".

The loosening of the traditional social structure, in these Arab villages, expecially the break-up of the rigid frame of the clan ("Hamula") has facilitated the mentioned changes in site and pattern of the villages. Until recently the population of most villages was divided into several clans. Each clan was confined to a clearly delimited part of the village. This division is absent on the new site or in the dispersed extensions of the old village.

The substantial increase in the financial resources of the average family in these villages has played a major role in the transformation of site and pattern. In the past the peasants in the area under consideration had been extremely poor. These marginal lands could only provide a very meagre existance, due to frequent droughts, poor shallow soils and primitive inefficient agricultural practices. In recent years, however, much money has been flowing into these villages from men who migrated to work in the rich oil principalities of the Persian Gulf. Further, many of the villagers are finding work outside the village, mainly as hired unskilled labour in public works, in urban housing projects and in urban services. Income from agriculture has also risen sharply. Much of the traditional subsistence agriculture has been replaced by market gardening and fruit plantations. A growing number of villagers could, thus, afford to build themselves new better located homes.

The rapid natural increase of the population of these villages (nearly 100% net over the last 30 years, despite emigration) has also been an important factor in the transformation of site and pattern and in the emergence of hamlets and ecarts.

The extent of change in pattern of the above mentioned villages can be expressed by the following simple formula:

$$D = \frac{A}{a} + \frac{H}{h} + \frac{P}{p} \, ,$$

D = index of dispersion, A = the present built-up area of the village. H = the total number of households, P = the total population of the village, a, h, p represent respectively, the area, the number of households and the population of the old clustered part of the village. Before the transformation processes started, when the entire population was confined to the old site D was equal to 1 (as A was equal to a, H to h etc.,) In the late 1950-ies D grew to 3-4, in the late 1960-ies to 6-7. It has now (1975) reached 10-12 in the most rapidly developing villages.

The relative importance of recent "birth" of hamlets and ecarts around the old village and on its lands can be presented by Demangeon's formula:

$$K = \frac{E \times N}{T}$$

K is the index of dispersion, E the population of hamlets and/or écarts, N the number of hamlets and/or écarts, T the total population of the commune (the chief settlement + hamlets and écarts). Before the mentioned processes of transformation K was equal to 0, now K = 4-5 for villages with 8-10 hamlets and écarts. The changes in the characteristics and landscape of the rural settlement in the mentioned semi-arid marginal zone are, however, much more extensive than those represented by the numeric expressions of these formulas.

204

Costa F.J., Dutt A.K.
(USA, Akron, Ohio)

COMPREHENSIVE URBAN PLANNING PARADAIGMS AND RESOURCES CONSTRAINTS

Study Hypothesis

The constraints imposed by limited resources have exercised a differential impact upon the development of urban planning paradaigms. Many of these planning paradaigms reflect the needs and potentialities of the resource base of the time and place of their genesis. Few have taken into account possible changes in energy and resource utilization patterns imposed by apparent resource shortages and environmental protection needs.

Study Methodology

The study methodology consists of the following elements:
1. identification of major comprehensive urban planning paradaigms and
2. development of a comparative framework to evaluate each with respect to resource utilization.

The paradaigms utilized include:
1. the Green Belt and Garden City concept;
2. the Ciudad Lineal (lineal city) concept;
3. Le Corbusier's Ville Radieuse concept;
4. Wright's Broadacres concept;
5. Socialist city planning concepts;
6. post-war British New Towns concept;
7. Doxiadis' grid city concept; and
8. Soleri's Arcopolis' concept.

Each of these paradaigms or concepts shall be examined and compared with regard to the following elements:
1. transportation and movement patterns;
2. land use and population density patterns;
3. spatial expansionism;

4. relationship of existing settlements to new settlements;

5. resource-base involvement; and

6. their appropriateness or validity in the present context of resource constraints.

Study Outcome

The study outcome is essentially a reevaluation of urban planning paradaigms in light of current and projected energy shortages. Preliminarily it can be observed that most of the popular paradaigms of recent planning history have been wasteful of land resources, fuels and other exhaustible materials must be either modified or discarded. A new alternative paradaigm must emphasize a cooperative use of land and facilities. Therefore, a new physical framework for planning needs to be developed to suit the requirements of the future energy and material shortages. The guidelines for such framework should consist of: 1. unfettered expansion possibility of CBD, 2. combination of fast-moving mass-transit and private automobile ownership, 3. neighborhoods designed in Perry's outlines, but suiting modern circumstances, 4. town and country style living with inter-connected recreation-cum-green area system with treelines at the edges, 5. factories and/or government offices to be located at one side of the city complex, and 6. possibilities for an unlimited growth of such a city, but when it grows into 500,000 or a million, separate it by mandating a green belt. An initial vision of such a framework is attached herewith.

Freitag U., Kamasundara B
(Thailand, Bangkok)

MAPPING THE POPULATION CHARACTERISTICS OF DEVE-
LOPING COUNTRIES: THAILAND AS AN
EXAMPLE

The declaration of 1974 the World Population Year raised a worldwide interest in the rapid increase, the regionally differentiated growth rate, the irregular distribution, and the unbalanced nutritional, medical and educational provision of the world population. The discussion centered on the different population characteristics of industrialized and developing countries and the influence of the population growth on the long term social and economic development.

The base of the evaluation and forecasting of the population characteristics were the various population censuses. In many Asian countries population enumerations were conducted around 1960 and again around 1970 in accordance with a United Nations recommendation to obtain comparable population data. In Thailand the sixth official census was carried out on 25 April 1960, the following census on 1 April 1970 providing the most reliable population data until now.

The census data were used in various ways to show the regional disparities of population characteristics. The best means to visualize the problems of population concentration as well as dispersal are maps. For this reason various population maps were compiled and printed, either detailed maps of Thailand as national entity or of Thailand within neighbouring Asian countries. Both types of maps supply important direct and indirect information for national or regional social and economic planning.

Population maps of Thailand were compiled by individual authors as well as by government departments.

The contributions of individuals are mainly small-scale maps of selected aspects of population to accompany problem-oriented papers in journals or research reports like R. Ng's paper on internal migration or N. Nagavajara's long-term variation analysis.

More detailed maps of more population characteristics were published by Thai government departments. Examples are the Statistical Atlas of Population 1960 and 1970 by the National Statistical Office and the National Resources Atlas of Thailand by the Royal Thai Survey Department. As its name suggests, the Statistical Atlas of Population consists of 25 choropleth maps at 1:5,000,000 which show individual statistical population data of the amphoes (counties) grouped into 4 -

5 classes. The topics include population density, sex ratio, age groups, fertility, literacy, and employment, separately for males and females. So far only the first volume with the 1960 census results was published in 1972. The National Resources Atlas of Thailand included in its 1972 edition only one population map, a density map of the amphoes in 1970 at 1:2,500,000 with 9 density classes and 8 size classes of 3 types of urban centers.

At present the Royal Thai Survey Department is compiling and printing the maps of a new revised edition of the National Resources Atlas of Thailand. The atlas section on human resources will contain 13 maps at the scales 1:2,500,000, 1:4,000,000 and 1:5,000,000. They will not only show isolated statistical data of the population census, but they will combine various correlated population characteristics in one map. Their number will not only consist of statistical choropleth maps, but also of qualitative maps. A careful selection of topics will provide the essential information on the distribution, natural and migrational development, medical and educational provision and employment of population for national planning purposes.

One of the fundamental maps of the atlas is the first population distribution map of Thailand at the large scale 1:2,500,000. This map was compiled in three stages: First, all settlements of the topographic map 1:50,000 - suveyed in 1969 and 1970 - were transferred to an exact map at 1:2,500,000, then the boundaries of the census districts were plotted into the same base, and finally the census data were mapped as squares (urban population) and dots (rural population). The compilation encountered various difficulties because of the instability of settlements of the shifting cultivation farmers, the inconstancy and inaccuracy of internal boundaries in many areas, the dubiosity of some census results in comparison to vital registration data, etc. These and other difficulties face most developing countries and restrict the use of map keys which were worked out for population maps of industrialized nations.

208

The population distribution map is a fundamental map not only of national, but also of regional planning. For this reason, the Royal Thai Survey Department suggested in 1961 in its first proposal of a Regional Economic Atlas of Asia to the 3rd United Nations Regional Cartographic Conference the inclusion of the topic population in the atlas. A population density map (at province level) of most Asian countries was prepared and submitted to the 6th Regional Cartographic Conference in 1970. At present, the Royal Thai Survey Department is compiling a population distribution map of Asia. It will show the population around 1970 in metropolitan and other areas by squares and dots. The compilation of this map encounters even more difficulties than any national map. The reduction of the map scale to 1:5,000,000 and the simplification of the map key will overcome some obstacles. The use of other maps of the Regional Economic Atlas will further facilitate the compilation of this map of the human resources and problems of Asia and the contrast of industrialized and developing countries.

Gardavský V.
(ČSSR, Praha)

THE GEOGRAPHY OF RECUPERATION IN CZECHOSLOVAKIA

Among the most outstanding features of contemporary civilization ranks the need of the population – in particular the urban population – for recuperation. In Czechoslovakia, the "second home" represents a weekend bungalow or a cottage, originally a permanent rural dwelling; though the two are markedly different, particularly architecturally, but also in setting, they have some common features. They are the private property of people permanently dwelling elsewhere and using them only for recreational purposes at week-ends, holidays or during vacation.

In 1971, there were registered in Czechoslovakia 166,246 individually owned recreational buildings, 156,402 of them in the Czech and 9,844 in the Slovak Socialist Republic. Extrapolating contemporary trends and the number of recreational houses under construction, it can be estimated that by 1976 their number will exceed 20,000. In this context it has to be noted that in the Czech Socialist Republic alone building allottments cover 126 square kilometres of agricultural area, and by adding the acreage covered by gardens this figure will rise fourfold. The territorial distribution of recreational houses was, up till the beginning of the fifties, extremely asymetrical. The cottages were clustered alongside rivers and man-made lakes and ponds, keeping close to railway lines linking these areas with urban ones. Nowadays the pattern is changing, becoming almost concentric, spreading out in all directions. There appears a sort of "recreational sphere", the size of which corresponds to its nucleus, the city. In its arrangement can be found zones of various recreational density and intensity, given e.g. by the number of second homes for a given area or the proportion of the part-time population as compared with the number of permanent inhabitants. This diversity is obviously given by natural conditions of the areas concerned (water, woodlands) and by their accessability by public and private means of transport. Of course, there occurs the needs of crossing the spheres of interest between individual towns, due to the density of urbanization in Czechoslovakia, but of decisive impact remain cities with 100,000 and more inhabitants: at least one third of their population, and in the warm seasons more than 40 per cent, spend their week-ends away from their permanent dwellings. The "recreational sphere" of Czechoslovakia's main cities features almost identical zoning. The greatest density and intensity is found in districts nearest to the city. E.g. Prague with two suburban districts (Prague East and Prague West) has the highest territorial concentration of second homes of all Czechoslovakia. In 1971, there were 28,841 privately owned second homes on this territory of 1699 square kilometres, and Brno with the district Metropolitan Brno (1324 square kilometres) follows with 10,877 such houses. In Slovakia, though the development was somewhat different, Bratislava with the district Metropolitan Bratislava (1607 square kilometres) having 2,582 second homes, represents the highest concentration. The further development of the "recreational sphere" depends on the size of the nucleus, which is most evident from the example of the Capital. The inhabitants of Prague have their second homes in all districts of the Czech Socialist Republic, with the exception of four districts (Opava, Ostrava, Karvina, Břeclav), although with increasing distance their proportion is falling sharply. The above-mentioned distribution of second homes, with a high concentration in the immediate vicinity of the urban area, which can be considered a consequence of earlier development, no doubt with steadily increasing car ownership, shorter working hours and constantly rising incomes, together with other factors, has led to the spreading out of recreational territories to their present pattern.

On the basis of this brief geographical survey of private short-term recuperation in Czechoslovakia, a further conclusion can be drawn: if we agree with I.P. Gerasimov, that city dwellers strive to regain their physical strength and mental faculties in a natural environment, then we find on the basis of the above quoted examples, that the environment in territories with a high concentration of private second homes is not fundamentally different from the environment which city dwellers wish to leave behind. Paradoxically, recuperation defeats itself, the process of recuperating devastates the originally attractive natural environment, changing it to such extent that it becomes quite unsuitable for the basic purposes of recuperation. In some areas, a new problem is rearing its head: the recreational destruction of natural environment, which further development will render increasingly serious.

GAMBI L.

(Italie - Milan)

CARTE DE L'HABITATION RURALE EN ITALIE

Ce document est une synthèse nécessaire de toutes les enquêtes et monographies sur l'habitation rurale en Italie, réalisées durant ces cinquante dernières années, pour la plupart dans le cadre d'une initiative prise en 1926 et poursuivie jusque vers le 1960 par Renato Biasutti, et en partie également par diverses lignes de recherche autonomes. les études entreprises et coordonnées par Biasutti (et continuées, après lui, jusqu'à nos jours) ont été presque exclusivement l'oeuvre de géographes; les autres recherches ont été faites particulièrement par des urbanistes, des économistes, des éthnographes, des historiens locaux. En 1950 il y avait moins de 500 ouvrages se référant à ce sujet; en 1970 - lorsqu'on édita par les soins de G.Barbieri et de L.Gambi un volume destiné à donner une vision globale des études faites jusqu'ici - ils avaient déjà dépassé le millier. Il manquait cependant à ce volume une carte de synthèse: et la carte ici élaborée a précisément l'intention de répondre à ce besoin d'une vue d'ensemble claire et minutieuse.

La carte qui a été dessinée montre les structures de l'habitation des travailleurs de la terre (au sens le plus large du mot) au milieu de notre siècle: à proprement parler, la situation entre la fin de la guerre et le milieu des années 60. Années qui sont - malgré l'impitoyable contradiction de nombre d'évènements et de processus - une période de transformations radicales dans la société agricole de notre pays. L'expansion, très importante dans certaines régions, de l'industrie et les formes massives d'urbanisme qui en découlent, entraînant des migrations internes de population, la déruralisation de zones agricoles de plaine et le dépeuplement de zones agricoles et pastorales de montagne; la spécialisation des cultures agricoles à des fins commerciales dans bon nombre de régions les plus fertiles; la crise mortelle du métayage qui conduit à sa disparition dans la plaine du Pô et dans de nombreuses contrées du nord de la péninsule; les réformes agraires qui englobent différentes zones, particulièrement le long de la côte de la péninsule et des îles etc. sont autant de phénomènes qui agissent énergiquement sur les structures des installations rurales. C'est pourquoi cette carte exprime une situation à la fois tourmentée et dynamique dans le contenu et la forme de l'habitation rurale en Italie; et les conditions que cette carte décrit ne sont plus celles qui existaient dans le premier quart du siècle, lorsque pour chaque région du pays l'habitation se prêtait à l'étude avec une claire stabilité de structures enracinées dans les traditions de plusieurs siècles. A ces conditions, qui dans certaines régions se retrouvent comme une véritable toile de fond du tableau, ou qui n'apparaissent plus avec la même incisivité qu'au début du siècle, nous en voyons maintenant d'autres se juxtaposer, avec des graduations de modes et de mesures qu'il n'a pas été aisé de représenter. Cette carte ne photographie donc pas une vue d'ensemble synchrone, pour ainsi dire, mais filme une réalité en mouvement.

Ce que je tiens en réalité à souligner, c'est que la carte établie ici se base sur des normes d'interprétation différentes des critères adoptés pour les études faites sur ce thème, il y a encore peu d'années, principalement par des géographes et des urbanistes: études dont j'ai utilisé les résultats plus comme base de documentation que pour la méthodologie de leurs schémas de catégories. A mon avis l'habitation rurale, dans les éléments qui déterminent sa composition et ses dimensions, dépend tout d'abord de rapports de production déterminés et d'une organisation agronomique déterminée. En un mot, il s'agit de l'expression de ces deux phénomènes sur le plan des aménagements. Et donc - comme le prouvent les évolutions en cours - cette configuration évolue sur un arc périodique plus ou moins long, suivant le changement des rapports et de l'organisation. Les autres éléments qui caractérisent l'habitation, au contraire, c'est-à-dire ceux qui sont dus aux conséquences inévitables des situations du milieu ou de la persistance de traditions éthnoculturelles (qui influencent toutes deux également la technologie de la construction), bien qu'on les enregistre partout et qu'elles apparaissent parfois remarquables sur le plan esthétique, sont moins significatifs pour une enquête sur les structures de l'habitation.

Conformément à une telle interprétation du sujet, le dessin de la carte s'articule sur deux plans: un plan en couleur indiquant les conditions de fond et un autre en noir indiquant les formes fonctionnelles de l'habitation. Pour finir, une ligne de couleur vive délimite les surfaces où les travaux de bonification hydraulique et agricole, ou la réalisation d'une réforme agraire, apportèrent durant ces cinquante dernières années les conditions nécessaires à une transformation - mûrie jusqu'à nos jours de façon fort diverse - des formes fonctionnelles transmises par la tradition. Plus précisement, on indique de différentes couleurs les rapports de production disparates (les économistes les appellent "types d'exploitation") qui dominent dans chaque région, et par divers signes en couleur - selon les types d'exploitation - on indique les cultures caractéristiques de chaque contrée .

A mon avis, cette carte peut particulièrement stimuler les reflexions de ceux qui s'adonnent à l'histoire dans une direction structuraliste. Parmi les faits qui ressortent tout d'abord clairement, je signale l'entière régionalité du tableau qui détruit toute hypothèse de vision unitaire; qui est plus marquée là où les pôles urbains d'une importante force économique et politique, promoteurs donc d'orientations évoluées ou d'évolutions particulières dans l'organisation agricole, ont été les plus nombreux; qui se découpe sur des plans différents des évènements historiques selon la nature des principes qui ont inspiré cette organisation au cours des cent-cinquante dernières années: c'est-à-dire le mercantilisme industriel dans la plaine du Pô, le mercantilisme préindustriel dans le centre de la péninsule, la féodalité des grandes propriétés foncières dans le Sud, y compris le Lazio et les îles.

On n'a même pas l'impression que la conséquence des affranchissements et des changements fonciers intervenus dans bon nombre de régions de plaine depuis le debut du siècle, et les processus d'urbanisation des espaces agricoles qui ont litteralement explosés autour des centres industriels au

cours des vingt-cinq dernières années, soient de nature à
avoir qielque peu modéré cette disparité régionale.

Hájek Z.
(ČSSR, Brno)

DEVELOPMENT OF THE EMPLOYMENT RATE IN THE
CZECH SOCIALIST REPUBLIC FROM SPATIAL ASPECT.

In contradistinction of many other countries the employment rate in Czechoslovakia exhibits a still rising tendency
after World War II. This holds both for the Slovak Socialist
Republic (SSR) and the Czech Socialist Republic (ČSR) where
the employment rate has held one of the leading positions in
Europe already since the beginning of this century. During
the period from 1961 to 1970 the employment rate increased
both absolutely and relatively on the territory of the ČSR.

1. Main indices of the increase of employment rate in
the ČSR

Index	Number of		% of employment rate	Total number		% of women employment rate	
Year	inhabitants	employees		of women	of this number employed	total	of the number of women
1961	9,571,350	4,695,264	49.06	4,930,814	2,017,876	42.98	40.93
1970	9,805,896	4,978,250	50.77	5,057,190	2,275,465	45.71	44.99
Index	102.45	10.603	x	102.56	112.77	x	x

The increase of the employment rate is in harmony with the
development of economics and was secured above all by a higher number of employed women (+ 12.77%). The percentage of employed women of the total number of women increased by 4.07%.
The ČSR belongs on the basis of both the total employment
rate of population (50.77%) and the women employment rate
(45.71%) to countries with a high employment rate.

2. Development of the employment rate in the ČSR according to spheres

Sphere	Absolutely			Relatively			
Year	I.	II.	III.	I.	II.	III.	Sa
1961 T	982,876	2,362,688	1,349,700	20.93	50.32	28.75	100.00
1961 W	519,127	823,603	675,146	25.73	40.81	33.46	100.00
1961 %	52.82	34.86	50.02	x	x	x	x
1970 T	736,326	2,460,525	1,781,399	14.79	49.43	35.78	100.00
1970 W	341,158	959,242	975,065	14.99	42.15	42.86	100.00
1970 %	46.33	38.98	54.74	x	x	x	x
Index T	74.92	104.14	131.98	T = altogether			
Index W	65.72	116.47	144.42	W = women of the total number			

According to economic spheres (primary, secondary and tertiary spheres) the development was as follows: In the primary sphere (agriculture) the employment rate decreases both absolutely and relatively and more quickly with women which is favourable with respect to the character of employment in this branch.

In the secondary sphere the employment rate is increasing absolutely, which is made possible by employment of a higher number of women. In relation to the total employment rate the employment rate of this sphere has a stable state.

The tertiary sphere (services) exhibits a considerable absolute and relative increase of employment rate. The increase of employed women increased by almost 50% (i=144.42) in the course of the period investigated, its percentage attaining 45.71%.

Special attention in the analysis of the employment rate from the territorial point of view should be paid to the course of the employment rate in urbanized regions. They differ more or less from other territory and can be characterized as regions with a progressive development of economics. A

Indices / Year	Share in the territory of the ČSR in %	Population number	Share in the population number of the ČSR in %	Number of employees	Share in the employment rate of the ČSR %	% of employed women of the number	
						of employees	women in the region
1961	11,15	T 3,784,000 W 1,963,358	39.53	T 1,840,872 W 780,003	39.21	42.37	39.73
1970	14.05	T 4,279,100 W 2,220,017	43.62	T 2,197,042 W 1,001,368	44.13	45.58	45.11
Index	130.08	T 113.08 W 113.07	x	T 119.35 W 128.38			

T = altogether
W = women of the total number

successive concentration of employment rate and population resulting in the development of urbanization take place here. They are characterized similarly as in other countries by more intensive dynamics of the economic life surpassing the usual average and display their specific character, and their independent position in the economics of each country.

3. Indices characterizing the urbanized regions in the ČSR

Urbanized regions are increasing. Thus even the share in the population and employment rate is increasing (by about 5%) besides immigration. The fact is substantial here, that the employment rate is increasing not only absolutely but even relatively and thus the values of 1970 are higher than the territorial average of the ČSR not only as to the employment rate of women (45,58%) but also to the total employment rate (44.13%).

From the spatial point of view the regions exhibit a relatively regular internal distribution of the employment rate and this regularity kept preserved even in the course of the period investigated. Highest concentration of employment rates occur in the core of regions with successive decrease towards the margins.

4. Spatial distribution of the employment rate in urbanized regions in %

Zone Year	Core	Urbanized territory	Marginal zone	Altogether
1961	83.90	11.45	4.65	100.00
1970	82.36	11.34	6.30	100.00

A relatively regular distribution of the employment rate in the regions can be found even from the aspect of economic spheres.

5. Employment rate according to spheres in urbanized regions

Relative in %

Sphere Year	I	II	III	Altogether
1961 1970	9.02 6.84	55.49 50.26	35.49 42.90	100.00 100.00

Absolute:

Index	90.39	108.09	144.33	119.35

The development of the employment rate according to the different spheres in the urbanized territories is analogous to the development in the whole ČSR. But the share of the various spheres is different showing that the development in the urbanized territories takes place in subject and time margin (relative and absolute decrease of the primary sphere, absolute increase of the secondary sphere in the case of optimum relative representation and relative and absolute increase of the tertiary sphere).

The studies resulted in the statements that in the ČSR:

- the employment rate displays a stable and slightly increasing tendency

- more progressive trend of employment rate occurs in urbanized regions

- the number of employees in the primary sphere decreases (agriculture), but their number increases in industry and mainly in services

- the total number of employed women increases

- in the future the employment rate will not decrease if the present population development keeps preserved.

Havrlant M.

(ČSSR, Ostrava)

SUR LES METHODES DE LA CLASSIFICATION DE LA CHARGE
MAXIMALE DES LOISIRS DE PLEIN AIR

La géographie actuelle du mouvement touristique dans sa plus grande partie est orientée vers la délimitation et ana-

lyse de la sphère tertière dans les rayons destinés aux loisirs de plein air, éventuellement vers l'étude des environs
plus éloignés convenables à la récréation de la population
de grands centres industriels. A cet égard est indispensable
l'étude de la région, de ces conditions de nature ainsi que
des influences socio-économiques facilitant la récréation
dans les possibilités de ces rayons.

La mesure de charge maximale de la région est fixée non
seulement par la dimension des rayons destinés à la récréation - représentés en Tchécoslovaquie surtout par les terrains boisés, mais aussi par la proportion des vacanciers et
celle des habitants locaux.

Pour établir le maximum de charge des loisirs de plein
air il faut tenir compte de relation entre le nombre des
visiteurs et celui des habitants locaux ainsi que de la corrélation de la fréquentation de l'endroit et de l'étendue du
terrain boisé dont on peut se servir jusqu'à la distance 4 -
5 km de l'emplacement des habitations - ce qui représente
1 - 2 h de marche à pied.

Jusqu'à maintenant on a pris pour la limite la plus
élevée de la relation réciproque avec les habitants locaux
la proportion 1 : 1, pour laquelle on se servait de termes
comme: l'index de la fonction touristique, la densité du mouvement touristique, le potentiel de la récréation. Il y a
des auteurs qui augmentent la mesure supportable du peuplement
des terrains boisés de 1 à 10 hommes sur 1 ha de leur étendue.

Un nouvel élément pour la classification de la charge
maximale sont les bâtiments destinés à la récréation, des
chalets particuliers ou des centres de détente des entreprises qu'on peut classer en même catégorie comme les hôtels et
des habitations pareilles. En Tchécoslovaquie le nombre des
bâtiments servant à la récréation en corrélation avec les
terrains boisés était jusqu'à maintenant fixé par rapport
d'un chalet au maximum à 5 ha du terrain boisé. Mais il faut
chercher un dénominateur commun pour toutes les habitations
destinées à la récréation et prendre en considération le nom-

bre des lits (la fréquentation), car celui de petits chalets
particuliers diffère considérablement de la capacité des
hôtels et établissements pareils. Voilà pourquoi dans mon
analyse de la sphère des loisirs de plein air appartenant à
l'agglomération industrielle d'Ostrava (c'est-à-dire les montagnes de Beskydy et Jeseniky) j'ai tâché d'en instituer une
nouvelle mesure designée par le terme "unité de chalet" pour
laquelle je prends un chalet particulier ou 10 lits d'un
établissement type hôtel - évaluation équivalante à 4 lits
d'un chalet privé en raison de la concentration de l'héberge-
ment, de l'utilisation de l'équipement commun, ce qui mène à
une exploitation plus intense de l'endroit.

Pour pouvoir évaluer la charge maximale de ce terrain
je me sers de corrélation entre les éléments du système correspondant:

1. L'étendue des terrains boisés destinés aux loisirs
de plein air à la portée des habitations en comptant avec
une distance diminuée de 20% au moins à cause des parties
impénétrables du terrain, des pentes abruptes, etc. Désormais
elle sera désignée par la lettre "P".

2. La fréquentation touristique précisée par le nombre
des lits de capacité du logement libre ou borné - la lettre
"L".

3. Le nombre des habitants locaux - la lettre "O".

4. Le nombre des unités de chalet c'est-à-dire un chalet
ou 10 lits dans un établissement type hôtel - la lettre "J".

En quelques variations on en obtient trois indices de
base: l'indice des constructions servant aux loisirs de
plein air   - "Z",
la densité du mouvement de loisirs de plein air   - "H",
l'intensité du mouvement de loisirs de plein air   - "I".

$$Z = \frac{J}{P} \times 5$$ La densité maximale des constructions des chalets est 1 chalet à 5 ha de la superficie
destinée aux loisirs de plein air ou 0,2 chalets (unités de chalet) à 1 ha.

$$H = \frac{L}{O} \times 1000$$ Pour la limite la plus élevée il faut prendre
d'un mille ce qui en corrélation 1 vacancier

avec 1 habitant représente des prétentions
très grandes aux services.

$$I = \frac{L}{P} \times 100$$

Pour la limite tolérable je prends 5 personnes
à 1 ha du terrain servant aux loisirs de plein
air. La limite souvent admise de 10 personnes
dans cette corrélation (les lits) il faut abais-
ser d'après moi à la moitié vu le mouvement
touristique, camping, excursions, écoles de
neige, etc.

D'après nos expériences un indice ne suffit point à la
classification infaillible de la charge maximale du mouvement des loisirs de plein air. Un grand nombre des chalets
concentrés à un seul lieu ne doit pas encore signaler une
superdimension du terrain, s'il y en a des surfaces assez
grandes ainsi qu'une fréquentation très élevée peut être en
accord avec des services, s'il est orienté vers des grandes
communes bien desservies. Pour pouvoir se rendre compte des
capacités libres des terrains pour les loisirs de plein air
de même que de l'organisation des services je prends en considération tous les critères cités et pour sa charge, je
prends un tel terrain dans lequel au moins deux indices dé-
passent la limite de la charge maximale même s'il faut prendre pour le facteur limitant le potentiel de loisirs de
plein air se rattachant aux terrains.

Sur la base des critères obtenus j'ai verifié cette méthode dans une région considérée comme typiquement surchargée par le mouvement des loisirs de plein air avec tous les
phénomènes secondaires négatifs.

D'après ce qui est montré ci-dessus on peut se rendre
compte de l'influence du nombre absolu des habitants ou de
l'étendue des terrains destinés aux loisirs sur les indices
avec lesquels les constructions et la fréquentation sont en
corrélation la plus étroite. Les données qui viennent d'être
citées démontrent qu'une valeur dépassée donne encore la possibilité d'exploitation continue, tandis qu'avec deux indices au-dessus de la limite supérieure nous risquons déjà de
déprécier la région.

La région destinée aux loisirs de plein air doit naturellement à cet égard remplir aussi les autres conditions,
telles que la densité plus basse du peuplement, la région
sans influence négative de l'industrie, l'accessibilité fa-
cile, l'équipement, les valeurs estétiques, etc. De ce point
de vue cette région est un géosystème dynamique dont la partie fonctionnelle est rélativement variable, tout de même
avec une limite supérieure. Celle-ci n'est pas donnée par
une valeur absolue mais relative au nombre des visiteurs et
aux éléments pris en considération en corrélations mentionnées.

La région des loisirs de plein air des montagnes de
Beskydy

la commune	le signe	Řeka	Moravka	Raškovice	Kunčice p.Ondř.	Rožnov p.Radh.
les fôrets (en ha)		1015	7139	75	731	1514
la superficie disponible pour les loisirs (en ha)	P	800	5700	50	600	1200
les chalets particuliers (nombre)		233	496	156	710	282
les bâtiments appartenant aux entreprises (h.d. lits)		178	840	15	747	180
les unités de chalet (nom.) le nombre des lits dans	J	251	580	157	785	300
les unités de chalet		1110	2824	639	3587	1308
la capacité libre du logement-nombre des lits		89	18	4	–	408
le nombre global des lits disponibles pour les lois.	L	1199	2842	643	3587	1716

la commune	le signe	Řeka	Moravka	Raškovice	Kunčice p. Ondř.	Rožnov p. Radh.
le nombre des habitants	O	493	2400	1547	2347	8503
l'indice des constructions	Z	1,56	0,5	15,7	6,54	1,25
l'intensité du mouvement des loisirs de plein air	I	149,7	49,8	1286	597,8	143
la densité du mouvement des loisirs de plein air	H	2432	1184,1	415,6	1528,3	201,7

IMBRIGHI G.

(Italy, Rome)

TERRITOIRE, TECHNOLOGIE ET PROBLEMES DE L'HABITAT

L'auteur de ce texte veut dire d'une manière succincte
où en sont arrivées les recherches qui ont été faites dans le
cadre de l'Université, plus précisément pendant les cours de
Technologie B de la Faculté d'Architecture de l'Université de
Rome, et également par les groupes de travail du secteur des
jeunes de l'UPTI (Union de la Presse touristique italienne)
faisant partie de la FNPI (Fédération Nationale de la Presse
italienne), secteur dont le signataire du présent article est
responsable. L'objectif de ces recherches (effectuées sur des
territoires-témoins dont nous avons déjà parlé) a été plus
précisément de clarifier et d'approfondir ce qui a été trai-
té dans deux textes précis. Le premier texte traite des rapports intercurrent entre la géographie du territoire et le
degré d'urbanisation de ce même territoire. Le second, des
rapports entre "activité" et "exigences", activité signifiant
les fonctions de résidence-production-loisirs, et exigences,

les qualités spatiales à l'échelle architecturale, urbaine et territoriale qui assurent l'activité.

En ce qui concerne le premier texte - rapports entre la géographie du territoire et degré d'urbanisation - il convient de dire que l'on a essayé de ne plus s'en tenir à la méthode traditionnelle pour analyser ces facteurs, c'est-à-dire en ne voyant que la simple dépendance pour les choix précis d'urbanisme, de facteurs géomorphologiques simples et complexes, où pire encore, en analysant séparément les deux moments, sans pour cela étudier les rapports qui peuvent exister entre eux.

Les domaines d'études, comme les recherches concernant l'origine et le développement des établissements spontanés, qui, sur une petite échelle, se concrétisent dans l'architecture indigène, ou encore la genèse des tissus urbains, que l'on peut comparer parfois, à la lecture des cartes, à la naissance et au développement des tissus physiologiques, offrent un nouvel intérêt.

Les domaines d'études dont nous venons de parler sont seulement des exemples des centres d'intérêt que ce premier domaine de recherche recouvre; il reste évident que ce qui nous intéresse le plus est de connaître la méthode de travail dont on s'est servi, mais cela n'entre plus dans le cadre de notre "short texte".

En ce qui concerne l'étude des activités et des exigences humaines, au sujet de leur influence sur le territoire, on devrait actuellement diriger la recherche non pas sur des éléments canoniques de base, comme un échantillonage d'activité humaine capable de représenter un habitat déterminé ou bien une série de ceux-ci, mais plutôt en partant de normes de base pour pouvoir parvenir à la définition des programmes des interventions à faire sur le territoire construit, comme celles qui ont déjà été faites mais que l'on doit accélérer pour sauvegarder opportunément le territoire naturel et les éco-systèmes correspondants.

Le problème évident qui se présente alors, et celui de la nécessité d'une quantification, au sens concret, des données au début de la recherche comme à la fin de la recherche.

Il faut ajouter que pour cette dernière partie des études on se sert de tissus-témoins de l'Italie centrale (et spécialement des régions Latium-Abruzzes), là où il est possible de trouver des zones de territoire au "degré d'urbanisation" très varié dû à des conditions bio-physiques et géo--morphologiques extrêmement différenciées et qui se prêtent donc bien à ces études.

Lleshi Q.
(Yougoslavie, Prichtina)

LE ROLE DU VILLE EN ORGANISATION DES ACTIVITES AGRICOLES (EXEMPLES DES PAYS SOCIALISTES EUROPEENS)

Le rôle organisateur de la ville vis-à-vis de la campagne environnante et sa production agricole est un nouveau phénomène apparu en pays de l'Europe de l'Est après la deuxième guerre mondiale comme conséquence du changement de leurs systèmes politiques et économiques. En effet, le rôle de la ville en commercialisation des produits agricoles reste une des fonctions principales, notamment des centres régionaux de toutes les villes nonobstant leur dimension. En plus, la fonction du ramassage et de la commercialisation des produits agricoles de la campagne environnante est une des fonctions initiales d'apparition et de développement urbain d'une agglomération c'est-à-dire les agglomérations en effet commencent leur première "baptême urbain" avec cette fonction. Ce chemin de développement urbain est notamment caractéristique pour les villes dans différents pays de l'Europe du Sud-Est.

Les changements principaux des fonctions des villes de l'Est (pays socialistes européens) vis-à-vis de la campagne environnante consistent en l'augmentation des fonctions nodules en cette sens. C'est-à-dire que exclusivement en système socialiste les villes conquièrent leur fonction organisatrice de la production agricole et généralement le rôle des transformateurs du paysage agraire environnant.

Il faut souligner que le rôle organisateur d'activité agraire des villes dans les pays socialistes est apparu récemment c.à.d. pendant les dernières 15-20 années, parallèlement avec le developpement et l'agrandissement des habitations urbaines et industrielles (augmentation adéquate de la demande sur le marché intérieur des produits alimentaires). Dans les pays typiques agraires de l'Europe de l'Est:

Albanie, Bulgarie, en partie aussi Hongrie, Roumanie, Yougoslavie, ce rôle est conditionné par l'exportation de leurs produits agricoles sur le marché de l'Europe centrale qui a besoin de telles marchandises que: légumes, raisins, vin, tabacs, fleurs, etc.

La première condition favorable pour cette fonction des villes dans les pays socialistes européens est créée après la deuxième guerre mondiale, avec les changements radicaux de la propriété foncière qui partout est devenue plus ou mois collective ou bien d'Etat.

C'est évident que ce rôle de la ville n'est pas le même dans chaque pays socialiste de l'Europe, c.à.d. que cette fonction est conditionnée par la réglementation législative agraire en chaque pays; il dépend aussi des conditions écologiques non seulement de chaque pays, mais, également des differentes régions dans un même pays. Les grandes différences sont notamment évidentes entre les pays avec une collectivisation totale de la terre (Albanie, l'Union Soviétique, Tchécoslovaquie) et les pays qui ont moins d'un quart de la superficie agraire collectivisée (Yougoslavie, Pologne).

Le cas de la Yougoslavie est différent de tous les autres pays socialistes de l'Europe. A l'exception des coopératives agricoles en quelques grands villages, toutes les grandes entreprises agricoles, le plus souvent mixtes, c.a.d. agro-industrielles (les combinats) sont localisées dans les villes. Une autre différence consiste en ce fait que tous les combinats et coopératives en Yougoslavie travaillent par le compte économique. Les interventions d'Etat, c.à.d. les subventions, avant la réforme économique très repandues, aujourd'hui sont très rares.

La localisation des combinats agricoles exclisivement en villes a sa raison purement économique et aussi sociale.

Les villes offrent les conditions économiques les plus favorables ainsi que la main-d'oeuvre nécessaire (ouvriers, cadres techniques, fonctionnaires). Les villes offrent aussi d'autres moyens nécessaires à servir la campagne: alimentation en matière première (farine, engrais), les moyens protecteurs des plantes, ainsi que les moyens de service technique (les ateliers mécaniques sont exclusivement en villes).

En terminant cette modeste étude on peut donner quelques conclusions générales:

1. Le rôle organisateur de la ville vis-à-vis de la campagne environnante du point de vue d'organisation de la production agricole dans les pays socialistes de l'Europe, a apparu après la dernière guerre mondiale.

2. Cette fonction a apparu dans la période d'après-guerre avec l'augmentation du marché intérieur des pays producteurs et en même temps avec l'augmentation de la demande du marché extérieur pour les denrées agricoles.

3. La création des entreprises agricoles ou mixtes (agro-industrielles) a lieu dans les villes moyennes (de 20 000 à 50 000 habitants).

4. Le rôle de la ville ne consiste pas seulement dans l'organisation de la production nette agricole, mais se manifeste en d'autres formes (travaux de la mise en valeur des terres friches, etc.).

Macka M.
(ČSSR, Brno)

CHANGES IN THE DEVELOPMENT OF COMMUTING IN THE CZECHOSLOVAK SOCIALIST REPUBLIC DURING 1961-1970

In the population census of March 1st, 1961 commuting to work was established on state scale in Czechoslovakia for the first time. The population census of December 1st, 1970 did it too. Both censuses give the possibility of judging the changes in commuting in the course of the period investigated.

The conceptions of the censuses are somewhat different as to methods. Both are based on the evidence of out-commuters according to residence and administrative structure and hierarchy. But in the census 1961 only 325 selected towns were watched as centres of commuting, in the census 1970 the number of centres investigated was 1920. In the census 1970 the factual distance of the residence and workplace of every in-commuter was registered as well as the time necessary for this trip. Since this point of view has no corresponding equivalent in 1961 conclusions of this kind can be made only by comparison of out-commuters across the borders of administrative structures of different orders.

The following changes took place in the period between 1961 and 1970:

1. The commuting increased absolutely and decreased relatively. In 1961 altogether 2, 324,000 people (from this number 1,670,000 in the Czech Socialist Republic and 654,000 in the Slovak Socialist Republic) commuted to work, i.e. 43.1% of economically active inhabitants (in the ČSR 40.0%, in the SSR 52.7%). In 1970 altogether 2,598,000 persons commuted to work (increase to 111.8% in comparison with 1961, from this number 1,748,000 in the ČSR and 850,000 in the SSR), i.e. 37.2% of economically active people in Czechoslovakia (35.1% in the ČSR and 42.4% in the SSR).

2. This development is in harmony both with the trend of the concentration of population in larger settlements as well as with successive disengagement of agricultural population. The third concomitant phenomenon is the successive concentration of small locally scattered manufactures into greater units and larger communities. This is documented by the data in the following table.

A distinct border form urban settlements with 5,000 up to 10,000 inhabitants. To this border both a decrease of population and a decrease of economically active population take place. As to commuting which is a resulting phenomenon the process of concentration manifests itself from the size group with 1,000 up to 2,000 inhabitants. Owing to the relatively low population number in cities the most dynamic group is the

229

Size group in th.	Population number in %		Increase decrease 1961-1970	Economically active population in %		Increase decrease 1961-1970	Out-commuters in %		Increase decrease 1961-1970	Share of out-commuters in economically active population in %	
	1961	1970		1961	1970		1961	1970		1961	1970
- 0.2	2.3	1.7	75.7	2.5	1.6	70.2	2.3	2.3	97.6	37.5	52.1
0.2-0.5	10.3	8.6	87.4	11.0	8.5	83.8	13.2	12.6	104.0	45.4	56.4
0.5-1.0	14.8	13.5	95.4	14.9	13.0	94.9	23.8	21.0	94.9	52.2	60.7
1.0-2.0	15.1	13.8	95.8	14.5	13.0	97.5	20.6	21.1	110.8	53.5	21.1
2.0-5.0	16.4	15.8	95.3	15.5	14.1	98.2	19.6	18.8	103.5	47.7	50.3
5.0-10.0	8.6	8.3	101.0	8.4	8.7	112.2	7.1	7.3	111.3	31.9	31.7
10.0-20.0	7.2	8.0	116.1	7.1	8.2	125.4	4.4	5.3	129.7	23.6	24.4
20.0-50.0	7.2	8.8	127.1	7.1	9.2	141.0	4.0	4.5	124.5	21.1	18.6
50.0-100.0	4.0	6.5	168.0	4.2	7.5	194.5	2.1	2.5	127.4	19.1	12.5
100.0+	14.1	15.8	117.4	14.8	16.2	118.1	2.7	4.6	203.1	6.9	10.8
Total	100.0	100.0	104.5	100.0	100.0	108.4	100.0	100.0	108.0	36.5	37.7

McDonald J.R.

(USA, Ipsilanti, Michigan)

LABOR MIGRANTS IN WESTERN EUROPE: PROBLEMS OF
REPATRIATION

Throughout the quarter-century following the end of
World War II, the movement of migratory workers from the Mediterranean basin to the industrial nations   farther north
(especially France, West Germany and Switzerland) was a significant and highly visible feature of the continent's population geography. At the beginning of 1974, some eight million migrant workers were employed in Western Europe; with
family members this represented nearly twelve million total
migrants. Economically, the largely unskilled migrants have
category of towns with 50,000 up to 100,000 inhabitants.

3. In the period investigated a considerable decrease of
distant commuting both generally and in the individual regions took place. This manifests itself most distinctly in
the SSR where the commuting of workers and employees decreased from 88,000 in 1961 to 48,500 in 1970. The most distinct
decrease of distant commuting appears in the town of Ostrava.

4. The concentration of commuting to the environments increased of workplaces. In 1970 only 12.2% persons commuted
out across the district border. In-commuting decreased mainly
from the marginal parts of the commuting regions of the various centres and transitional zones the width of which also
decreased in several places.

5. A new phenomenon, such as commuting between urban
centres in opposite directions increased considerably. In
cities it is just this kind of movement causing the greatest
increase of in-commuting in the period investigated. Inter-
urban commuting shared already in 1970 more than 1/3 in
total commuting and is still increasing. In eighties it will
become the prevailing type of commuting and will dominate
over commuting from small communities in the hinterland of
the centres.

customarily worked in the low-prestige, low-pay jobs largely
abandoned by local labor. Socially, they have aggravated al-
ready-acute urban problems and have become the objects of
various overt and covert forms of discrimination. The current
economic recession, which began to be felt in Europe in 1973,
has had a profound effect on the thinking of industrialized
countries where unemployment is rising, and may in fact signal the end of migrant arrivals in important numbers. It is
ironic that, at the very moment when social justice for the
migrant began to be a significant theme in most countries
with large migrant populations, highest priority must sud-
denly be given to the problems of adjustment which will face
the workers upon their repatriation.

Nováková-Hřibova B.
(ČSSR, Brno)

CAUSES OF MIGRATION IN CZECHOSLOVAKIA FOR THE PERIOD BETWEEN 1966 AND 1970

The analysis of migration allows to get to know the social-economic conditions of a certain area. Of course, besides the extent of migration its motives and relations to the
other elements of environment must be studied.

The Czechoslovak Statistical Board investigating the extent of migration since 1954 began to establish its causes in
1965. The migration motives were sorted both according to the
view-points of economic character, such as: change of workplace, getting nearer to the workplace, apprenticeship or
study, and according to viewpoints of family (social) character, such as: marriage, divorce, state of health and finally
housing problems. Besides these 7 reasons there exists a group
of "other reasons" without precise specification. This cir-
cumstance should be taken into consideration with respect to
the fact that only one i.e. the main motive of migration is
established.

On the basis of the analysis of the migration motives
their share can be quantified with regard to territorial

units of different size considering the sex, age structure
and the socioprofessional structure.

The period between 1966 and 1970 was characterized by
extensive migrations in Čzechoslovakia; in annual average
367,800 persons changed the place of residence. Most attractive as to migration is the North-Moravian Region and all
cities.

The cause of migration mentioned most frequently is "the
housing problem" (45.4% of all migrants in the country, 44.2%
in the Čzech Socialist Republic - ČSR and 46.3% in the Slovak
Socialist Republic - SSR). There is a relatively free dependence between the number of inhabitants putting forward housing problems as migration motive and the intensity of housing
construction.

The second most frequent reason for migration is "the
change of workplace" - (20.1%) its share being equal in both
parts of Čzechoslovakia. Closely linked to the change of the
workplace is also the reason of "getting nearer to the workplace" which holds the 4th place in the sequence occurring
mainly in Slovak Regions and connected closely with commuting
to work and its decrease.

The third most frequently alleged reason for migration
is "the marriage" (15.2% of migrants) relatively equally represented on the whole territory of Čzechoslovakia. In the
period between 1966 and 1970 together 610,000 of marriages
were contracted so that from 100 marriages the share of moving couples was 20 (CSR 19.9%, SSR 20.1%).

About 2.0% of migrants put forward "divorce" as motive
of migration. In the period investigated 110,600 divorces were
consented to so that from 100 divorcing couples the share of
migrants was 13.8 (ČSR 13.9, SSR 13.6).

Reasons of health as a motive of migration are put forward by 4.2% of inhabitants. A certain correlation in distribution to the state of environment can be observed.

Only 0.7% of migrants allege "apprenticeship or study"
as the reason for the change of their place of residence.

An interesting image of changes in the share of the main

233

reasons for in-migration or out-migration can be established
if they are investigated according to size groups of communities. The change of the workplace in the case of in-migrants
is relatively most frequent if communities with population
number below 2,000 persons are concerned, in the case of out-
migrants in the group of communities with 5,000 up to 50,000
inhabitants. The reason of "getting nearer to the workplace"
for in-migration increases with the size group of the community decreasing, on the contrary, in the case of out-migration.
The migrants putting forward as reason of migration "marriage"
or "divorce" cause the depopulation of communities with up to
5,000 inhabitants in favour of communities with more than
10,000 inhabitants. The population balance giving as reason
for the change of residence the state of health is characterized by a population decrease in the lowermost size groups of
communities with a population increase in communities with
more than 10,000 inhabitants, probably owing to more concentrated public health welfare.

The main cause of migration i.e. that of getting a flat
is characterized by a population decrease in smaller communities up to 5,000 inhabitants and a population increase in
communities with more than 10,000 inhabitants with concentrated housing construction.

The establishment of the correlation of population migration between the Czech Socialist Republic and the Slovak
Socialist Republic is of importance too. Bohemia and Moravia
have a population increase from Slovakia mainly owing to the
change of workplace less owing to housing problems and marriage. A similar situation is in Slovakia.

Finally, it can be said that the period between 1966
and 1970 was characterized by a strong migration, depopulation of communities up to 2,000 and/or 5,000 inhabitants
mainly owing to the change of workplace or owing to housing
problems as well as by a constant population increase in big
towns.

Peris Persi
(Italy, Fano)

HISTORICAL CENTRES OF CITIES IN ITALY: EVOLUTION OF A CONCEPT OF VALUE AND FIELD OF GEOGRAPHICAL INTERVENTION

Every town centre derives from the fusion of geographical and historical factors which come together and form physical and man-made structures. The aim is to exercise certain functions better, among which one of the most important is the organization of the surrounding territory with which a phenomenon of resonance is established. It is a matter of dynamical functions which evolve according to human choices and are reflected in the architecture of the town centre. Sometimes these functions make the architecture adapt to new demands but more often bring it to a reduction, if not to a total extinction.

The traces and evidence which survived however, have badly become a part of the town and regional context which not only do not want them but do not even respect them. From here starts the degradation of old quarters or of whole centres. We have: a "human" degradation as these quarters or centres are abandoned by the population in search of more comfortable homes; an "ambient" degradation as the physical elements are left in decay awaiting to replace them with new constructions in consideration of the high profit of these central areas. The problem arose only in the second half of the nineteen fifties. Previously the social, political and moral crisis followed by the Second World War and by the pressing necessity of rebuilding had not allowed to see the true dimensions and articulations of the problem. In fact up to the fifties it had been a problem of single monuments which had to be isolated perhaps by thinning out the surrounding town texture. Only then did the concept of historical centre as unity and totality of situations come to be - under this aspect it was object of a series of studies which were

concluded in a national meeting which discussed the meaning-
ful theme: "Protection and Clearance of Historical and Artistic Centres" (Gubbio, 1960). This meeting represents a fundamental reference point not only because it brought to the
forming of "ANCSA" (National Association for Historical and
Artistic Centres), but also because it determinated a series
of interventions, meetings, discussions of national bodies
(such as "Italia Nostra") and of city administrations which
were the foundations for a broader discussion on a European
level.

In fact the nineteen sixties marked a period of studies
and of mutual explanation which were followed by bills brought
before Parliament above all in occasion of ANCSA conventions
but which were systematically ignored by it. Then an esthetic
factor on one hand and a cultural one on the other prevailed.
An historical centre represented an historical and architec-
tonic value and as such it was endowed with a "sacredeness",
precious for everyone. Therefore it was neither to be abandoned nor modified to adapt it to the demands of an urban civilization, motorized and chaotic; it was to be conserved as
a sign of past generations.

At the same time the problem, although it was seen only
in these terms, assumed a new spatial dimension as it spread
out towards the thick web of small and very small centres
which have dotted the Italian landscape for centuries and
which had to be brought back to life. Therefore we went from
the stage where an historical centre equals an object of con-
templation: a simply conservative restoring is no longer sufficient; new functions which can reconcile themselves with
old and austere buildings are necessary.

The best thing to do seemed to make these centres
(and they did) the seat for schools, universities, libraries,
theatres, offices, clubs. A strange type of city derived from
this: a city alive only during the day, a city which was a
compromise between a monumental supermarket and a cultural
unit which had to undergo a strong human pressure daily (and
this pressure was not sufficiently relieved by making these

centres reserved only for pedestrians). Therefore the characteristics of the "museum-town", "sanctuary-town", "university-
town", "administrative-town" began to show (Ascoli Piceno,
Ravenna, Assisi, Urbino, Pisa, Lucca, Roma, Milan etc.). These
centres became artificial and were no longer part of the rest
of the town community.

The content of historical centre had to be therefore examined and discussed all over again. The conclusion was that
it could be protected by leaving, in it, its original social
stratum, which is its natural heir from preceding generations
rather than by putting new functions which would cause con-
gestion, excessive specialization and at times continuous human alternations. This was not in contrast but had the same
aim as the following evolutions of the content of historical
centre that is as an "economical good" (rather than just cultural) where there were buildings and services (which had to
be no doubt improved, but which were already existing) which
were not to be nor could be wasted at a time when house policy was in crisis. Therefore its recovery became necessary
and at the same time it would be an historical, artistic, ar-
chitectonic, functional and social recovery. It is also a territorial recovery, as whole territories which had been forgot-
ten for years are directly interested and became part of the
regional, economic and political reality once again.

Naturally, being such a general principle, the solutions
which can be given vary from case to case and can emerge from
a careful historical and geographical study which can investigate on the part the centre has the mutual existing relations, the level of hierarchy that has come to be, the possibility of inserting a metropolitan, integrated system.

The regional decentralization, opposite the inertness of
the central government, seems to have better conditions for an
effective and quick intervention. It can quicken the preparation of the detailed town-plannings of historical centres
which are part of the general town planning of single towns
and of the regional planning. In these centres residential

and productive installation must reconcile with respect for
the ambient in its historical and landscape expressions and
achieve a balance between going to far ahead (industrialization for consumption) and anachronistic, decaying ultra-
conservatism. In this delicate level of knowing and planning
there are many possibilities for a geographer in collabora-
tion with other specialists of this field. Although at present
such team-work does not work or at least it is not properly
organized and although Italian geographers have shown a greater interest in urban phenomenon in its reality and its prospects, it is time for a more active participation in studies
and research on geographical problems. In the first place
there will have to be a census of the historical centres considering their patrimony of living space, their formal and
functional aspects and their territorial location. A first
contribution can be given by the inventory which was begun by
the Ministry of Education in the picture of the protection of
the European cultural patrimony (I.P.C.E.). Here we can pick
out the essential elements to face a valid regional policy
which can bring to an active recovery of the historical centre;
a recovery which can be seen under four aspects:

(a) _conservative_, naturally together with the surrounding
territory as it is offen necessary to save the whole landscape
around the centre;

(b) _economical_, in order to reabilitate, to a residential
function, buildings and houses already existing which can be
modernly restored;

(c) _social_, to assure a continuity of human presence
therefore to assure the continuity of "life in the quarter";

(d) _functional_, in order to look for and restore the natural vocations which the centre is called to in a new and
more integrated territorial order.

Roucloux J.C.

(BELGIQUE, Liège)

LES FONCTIONS SPECIFIQUES DES PLACES CENTRALES DES NIVEAUX INFERIEURS ET MOYEN DE L'ORGANISATION URBAINE

LE CAS DE LA WALLONIE DU NORD-OUEST (BELGIQUE)

A chaque niveau de la hiérarchie urbaine est lié un certain nombre de fonctions centrales spécifiques.

La recherche de ces fonctions spécifiques présente un double intérêt. D'abord un intérêt théorique puisque l'établissement d'une hiérarchie des fonctions correspondant à la hiérarchie des centres permet de vérifier une des principales propositions de la théorie des places centrales. Ensuite, l'intérêt d'une telle recherche est pratique puisqu'il est alors possible de déterminer le niveau hiérarchique d'un centre par la seule connaissance des fonctions spécifiques qu'il exerce.

Le procédé que nous proposons en vue d'établir une liste des fonctions spécifiques des divers niveaux hiérarchiques a été testé lors d'une étude que nous avons consacré au réseau urbain de la Wallonie du Nord-Ouest . Les résultats obtenus ne concernent donc que la Belgique et seulement les quatre niveaux de base de la hiérarchie urbaine. Le procédé comprend trois phases différentes: une phase de collecte, une phase de classement et enfin une phase de triage.

1. La collecte des données

Il s'agit d'établir un relevé aussi complet que possible de toutes les fonctions centrales des centres de services de la région étudiée. La fonction centrale est une fonction tertiaire typiquement urbaine dont la particularité est de provoquer le déplacement des consommateurs, c'est-à-dire de posséder un certain pouvoir de rayonnement. Le centre de services ou place centrale correspond au territoire sur lequel sont rassemblées un minimum de deux ou trois fonctions centrales.

2. Le classement des données

Le classement des données donne lieu à l'élaboration d'une matrice mettant en relation les fonctions centrales et les places centrales du territoire étudié. Cette matrice est représentée par un diagramme qui, très simplement, donne la composition fonctionnelle précise de chaque place centrale. En abscisse, les centres sont classés selon leur centralité décroissante, c'est-à-dire selon le nombre de fonctions centrales différentes qu'ils exercent. En ordonnée, les fonctions centrales sont classées selon leur banalité décroissante, c'est-à-dire selon le nombre de places centrales où la fonction eśt exercée. Sur le diagramme, chaque place céntrale est donc représentée par une colonne sur laquelle la position en ordonnée d'une série de carrés noirs indique très simplement les fonctions centrales exercées par la place centrale. Ce type de diagramme donne une vue globale de la structure comparée des fonctions centrales des centres d'une région. Il met également en évidence la liaison existant entre centralité et banalité. Cette liaison ne se marque pas par paliers comme le voudrait le système théorique de W.Christaller (1933) mais plutôt d'une façon continue par une augmentation de la densité des carrés noirs à proximité des deux axes du diagramme. A chaque niveau de la hiérarchie ne correspond donc pas une série bien précise de fonctions centrales puisque la composition fonctionnelle des centres du niveau considéré est loin d'être constante. De ce fait, il est impossible de déterminer les fonctions spécifiques d'un niveau hiérarchique par le seul classement des données sous la forme du diagramme que nous proposons.

3. Le repérage des fonctions spécifiques

Le diagramme donne avec précision non seulement la structure des fonctions centrales des divers centres mais aussi la liste des places centrales où sont représentées chacune des 267 fonctions centrales relevées en Wallonie du Nord--Ouest.
De cette façon, on peut connaître à chaque niveau de la hiérarchie le nombre de centres éxerçánt chaque fonction centra-

le. En d'autres termes, il s'agit d'établir la banalité des fonctions centrales au sein d'un seul niveau de l'organisation urbaine.

Cette décomposition de la banalité totale en banalités par niveau est à la base du procédé permettant de reconnaître les fonctions spécifiques. En effet, nous avons adopté le principe suivant: une fonction est spécifique d'un niveau hiérarchique lorsqu'elle est exercée par plus de la moitié des centres du niveau et qu'elle n'est pas déjà spécifique du ou des niveaux inférieurs.

De ce principe très simple découle logiquement la suite du procédé de recherche des fonctions spécifiques. Il s'agit d'abord d'établir les banalités par niveau de chaque fonction centrale, de la fonction la plus banale à la fonction la plus rare, et ensuite, sur la base de ces banalités partielles, il s'agit de déterminer le niveau spécifique de chacune de ces fonctions centrales. Si l'on envisage les niveaux à partir de la base du système hiérarchique, on peut considérer qu'une fonction est spécifique du niveau à partir duquel les banalités partielles de la fonction deviennent constamment supérieures à la moitié des banalités partielles maxima attachées aux divers niveaux de l'organisation urbaine.

Suivant la méthode que nous venons de décrire brièvement, il est possible de repérer les fonctions centrales spécifiques des niveaux de la hiérarchie urbaine. Il s'agit là de fonctions hiérarchisantes ou ubiquitaires. L'analyse fait également apparaître des fonctions indifférentes ou sporadiques, distribuées sans considération de la centralité des places centrales et donc non spécifiques.

4. L'analyse d'un cas concret. Les principales fonctions spécifiques relevées en Wallonie du Nord-Ouest

Sur 267 functions centrales différentes relevées en Wallonie du Nord-Ouest, 177 sont spécifiques d'un des quatre niveaux de base de l'organisation urbaine. Les autres fonctions sont soit spécifiques d'un niveau supérieur (bassin de natation, commerce de parapluies), soit sporadiques (commerce d'antiquités, école agricole, café-dancing).

Huit functions sont spécifiques des centres élémentaires qui constituent le niveau de base de la hiérarchie: café, alimentation générale, boucherie, distribution d'essence, lieu du culte catholique, école primaire officielle, etc. Dix-neuf fonctions sont spécifiques du second niveau, celui des villages-centres: salon de coiffure pour dames, carrosserie-garage, agent d'assurances, boulangerie, cabinet de médecine générale, entrepreneur de peinture, pharmacie, bureau de poste, droguerie, commerce de chaussures, etc. Quarante-quatre fonctions sont spécifiques des centres locaux: magasin de meubles, librairie, architecte, marchand de journaux, maroquinerie, dentiste, vétérinaire, tailleur, notaire, fleuriste, photographe, gendarmerie, etc. Cent six fonctions sont spécifiques du niveau moyen de la hiérarchie, celui des petites-villes : restaurant, cordonnerie, taxi, graineterie, parfumerie, avocat, pédicure, infirmière, crêmerie-fromagerie, cinéma, enseignement moyen, marché forain hebdomadaire, salon-lavoir, pompiers, clinique, gynécologue, agence de banque, agence de voyages, agent de change, etc.

Ševčik František
(ČSSR, Olomouc)

LE TRANSPORT ET SON INFLUENCE SUR LE STYLE DE VIE
DE LA POPULATION

A présent le transport - l'une des disciplines de géographie économique - attire l'attention des spécialistes en géographie économique. Dans ma thèse "Le transport dans le district d'Olomouc et son rapport avec le transport dans la ville", j'ai tracé l'évolution et la situation actuelle du transport dans le district d'Olomouc. Evidemment, cette région représente une entité géographique assez petite. Les études régionales modernes traitent pour la plupart les régions en complexe ou bien elles se réduisent à l'analyse des phénomènes économiques ou naturels. Aucune de ces méthodes de géographie régionale ne peut être exercée quant au district d'Olomouc et c'est pourquoi l'auteur était obligé de

242

tracer les conditions économiques et naturelles du point de vue qui surpasse la frontière administrative du district en question.

Pour faire mieux comprendre l'importance économique du transport, il faut souligner son caractère singulier: le transport comme l'industrie, l'agriculture et l'industrie du bâtiment représente dans l'économie nationale un secteur productif indépendant, caractérisé par K. Marx dans sa Théorie de plus-value comme une branche de production indépendante.

Le transport diffère de l'industrie par quelques particularités qu'on doit respecter en jugeant la place du transport dans l'économie nationale.

Comme la circulation des marchandises entre les entreprises et les établissements, entre l'industrie et l'agriculture, entre la production et la consommation, entre la ville et la campagne est réalisée par les moyens de transport, chaque irrégularité du transport provoque des troubles dans le processus économique national.

C'est pourquoi la tâche et l'objet principal de la géographie de transport est l'analyse des rapports entre la distribution des centres de transport et de ses catégories du point de vue géographique et l'économie des régions en question. A côté de cela, la géographie de transport étudie aussi l'influence des conditions naturelles sur la distribution des centres de transport et leurs moyens; elle s'intéresse aussi aux causes de la distribution des centres de transport du point de vue historique.

Si nous parlons des moyens de transport en commun dans les villes, nous sommes obligés à la fois de nous occuper de la question du transport automobile individuel. Sommairement, je veux noter quelques aperçus sur son développement et sa situation actuelle. Dans la dernière dizaine d'années, le transport automobile individuel en Tchécoslovaquie a noté une évolution considérable. On suppose qu'en 1980 chaque famille sur deux va posséder sa voiture.

Cette évolution rapide du transport et son amélioration constante apportent à la fois beaucoup de problèmes difficiles à résoudre. Ces derniers empêchent la réalisation des

243

postulats cardinaux dans le domaine des transports, comme
p.ex.la sécurité, la vitesse, la commodité et surtout la con-
tinuité du transport. C'est la discontinuité du transport qui
provoque la plupart des collisions. Dans les grandes villes,
on veut résoudre les problèmes de transport par la recomman-
dation des transports en commun qui doivent remplacer les
moyens de transport individuels, etc.

Le développement rapide de l'automobilisme individuel
augmente aussi la pollution du milieu non seulement dans les
villes mais aussi dans les régions de récréation. Et c'est
pourquoi la circulation des moyens de communication individuels (des autos) dans les centres de grandes villes (Prague,
Rome, etc.) est limitée ou interdite et les habitants sont
obligés de se servir des moyens de transport en commun - ils
doivent quitter leurs voitures à la banlieue où sont à leur
disposition de grands parkings.

La ville d'Olomouc appartient aux 46 villes en Tchécoslovaquie 1971 qui donnent à leurs habitants les moyens de
transport en commun (les autobus et les tramways). En 1971,
on a transporté 46 millions de passagers - un habitant s'est
servi 575 fois du moyen de transport en commun. La coordination de ces deux moyens de transport en commun est très bon-
ne. Au contraire, la coordination du transport ferroviaire
et du transport urbain à l'aide des tramways est assez difficile et désavantageuse pour les tramways.

Le transport urbain individuel doit être jugé du point
de vue du transport urbain en commun. Comme le transport individuel se réalise sur les mêmes voies que le transport en
commun, la capacité des voies est surchargée et la circulation ralentie. L'abaissement de la vitesse de circulation
est causé par les tramways qu'on ne peut pas doubler pendant
leur arrêt à la station.

La géographie de transport devrait prêter l'attention
aux questions comme l'influence de l'automobilisme sur le
style de vie de la population pour qu'elle puisse préparer
les conditions au développement économique et culturel, ci-
inlus le domaine des loisirs, de notre société socialiste.

A cette fin, les specialistes en géographie économique doivent analyser la situation dans de petites régions de ville et de campagne et révéler en même temps les différences entre les régions plus ou moins industrialisées.

L'analyse en question doit respecter le caractère naturel et économique de la région étudiée, elle doit mettre en relief la structure sociale de la population et sa propriété des autos, l'emploi des voitures pendant tous les jours de la semaine pour aller aux bureaux ou aux usines, pour faire des achats, pour aller aux cinémas ou aux théâtres, pour faire du sport ou l'emploi des voitures dans les jours du week-end pour les voyages à la campagne, aux centres de loisirs ou l'emploi des autos pendant les grandes vacances pour voyager à travers le pays ou à l'étranger.

Une telle analyse fera base pour se faire l'idée sur le style de vie des diverses couches sociales de la population tchécoslovaque et pour les directives de travail des comités nationaux, du gouvernement de la République Socialiste Tchèque ou de la République Socialiste Slovaque.

Cette recherche pourrait nous aider à améliorer l'approvisionnement dans les régions de loisirs, à édifier de nouveux parkings et à mieux distribuer le réseau des postes d'essence.

Les recherches concernant le style et le niveau de vie de la population en corrélation avec le développement de l'automobilisme peuvent résoudre les questions se rapportant au milieu et à la protection de la nature qui sont au premier plan de l'attention du gouvernement.

La solution des problèmes dont nous avons parlé pourrait éclaircir les cohérences intérieures qui sont en relations étroites avec le développement de la société socialiste.

Sipka E.

(USSR, Trnava)

THE PROBLEMS OF SEASON DEVIATIONS OF LAKE
DISTRICTS TOURISTS TRAFFIC

Exemplified by Orava Dam in Slovak Socialist Republic

The building of Orava Dam was completed as a part of the process of Slovakia socialist industrialization as the highest degree of the river Vah cascade. As a result of its completion, the water surface of 3600 ha in the shape of the wide U has been formed, covering the bottom of Orava Hollow at the junction of the White and Black Oravas. The water surface assumed its recreation function, besides the former energetic, industrial, cleaning and hydrological ones.

The main factor affecting tourist traffic at the dam lake is first of all a complex of climatic elements, mainly temperature, which curve has the shape of Gaus's curve. In the area there are 240-260 days with the temperature above $0^{o}C$ (lasting from 10th to 20th March till 20th to 25th November), 144 days with the temperature above $10^{o}C$ (lasting from 1st to 11th May till 21st September to 1st October).

Summer days of an average daily temperature of $15^{o}C$ and more last since June 1st till August 11th. Temperature indications in Orava Dam region are the lowest of those of Slovak hollows. If the tourist traffic depended on temperature, it would have to be limited to the period from July 1st till August 11th. A much more important factor than temperature may be the sunshine period here, amounting to 1850-1950 hours (that is 40.2% out of the possible 4473 hours). Though these are the lowest values in Slovak hollows and only 1250 of them fall on summer season, 600-700 hours on winter season, even this factor must be taken into account. There are 150 cloudy days at Orava Dam, it rains for 120-125 days out of total, most precipitation falls during summer season in the form of short stormy rains, which do not represent a serious traffic obstacle, as lasting raining does not occur.

In the chart of time sequence of positive average climatic elements values (temperatures, sunshine, converted values of total daily rainfall), periods of average suitability for visiting Orava Dam will appear, occurring in a much wider time interval (from May 8th to October 3rd). They show that Orava Dam is a suitable tourist traffic region in a wide time scale in spite of the minimum of positive climatic elements.

The beginning and end of vegetation season are important from the point of view of tourist traffic. Its beginning is symbolised by hazel blossoming and moves about March 21st (usually from March 5th to April 17th), followed by agricultural work in 5-10 days as a rule, and it culminates by apple-tree blossoming about May 9th. The end of vegetation season is connected with succession of average daily temperature 5°C in Orava Dam region - it is about October 27th. The period from March 21st or May 9th can be considered to be fully attractive for tourism from the point of view of vegetation.

The knowledge of water fauna (mainly fish) life-cycle and hunting seasons, is also very important for tourist traffic; hunting season lasting since June 15th till water-surface freezing, as a result of it tourist seasons are prolonged till the end of December.

Besides the Orava Dam lake region itself, a very important factor (though a subsidiary one) is the character of adjacent country regions. Orava Dam is surrounded by 2 types of country regions: (a) countries with the more expressive physical-geographical sphere (Slovenské Beskydy, Oravska Magura, Skorušina and West Tatras); (b) countrysides with the more expressive social-economic sphere (the Sub-Beskydian Highlands, the southern part of Orava Hollow and Orava Valley). Countries with the more expressive physical-geographical sphere are potentially very precious. In tourist traffic they can affect the enrichment and as a result of it prolongation of the stay of Orava Dam visitors during the high season, mainly in its marginal periods (spring, autumn), as well as the stay of tourists at Orava Dam in general, during a winter season. Orava Dam provides a suitable place for recreation stay and near mountains (mainly Oravska Magura) enable the development of winter sports. There are quite a lot of frosty days (100-110) and icy days (45-50), a high total of winter rainfall (220-400) and mainly a long period with a permanent snow covering (from December 25th to February 23rd), the average maximum snow depth being 40-60 cm. However, these districts are not all equipped.

Areas with the more expressive social-economic sphere with their potential conditions all the year round can be used in Orava Dam tourist traffic mainly in marginal periods of the high season and at the time of occurrence of less favourable climatic elements during the high season.

Material-technical basis of Orava Dam tourist traffic consists of accommodation facilities, on the one hand, lending-offices of sport requisite and regular shipping traffic, on the other. Accommodation facilities distribution is determined by road course in Orava Dam region, following its south and south-west banks and by coast geomorphology, from which the more suitable section seems to be the south-west coast, creating the Orava Magura slope. That is why all Orava Dam accommodation facilities have been distributed here.

The total number of Orava Dam accommodation facilities is 19 of 1858 beds. Out of the total number of beds, 570 ones, i.e. 30,7% are permanent and 1288, i.e. 69,3% are seasonal. On this ground 395 640 overnight stays in all can be presupposed at Orava Dam, out of which 208 040 on the permanent beds, i.e. 41,8%, and 187 600 on the seasonal beds, i.e. 27.4%. A real number of overnight stays in Orava Dam region was 124 868, 86 906 ones falling on permanent accommodation facilities, i.e. 69,6% and 37 962, i.e. 30,4% falling on seasonal accommodation facilities; so the share of overnight stays in permanent accommodation facilities exceeded the number of bed-days, that is to say, these facilities were better used. The utilization percentage grows by Gaus's curve, this one not being too steep, as the tourist traffic is small besides July and August. That is why the main factor stimulating visits to Orava Dam is water surface; from the point of view of time,

248

the main localizing factor is most favourable climatic conditions (particularly those of temperature) other factors are school holiday, holiday traditionally planned for July and August, and youth and adults recreation. Summer holiday stay is the main form of stay Orava Dam.

Other possibilities of stay are not utilized, e.g. organizing outdoor school for children of regions infested with exhalations and big towns with the possibilities of an active recreation and informative excursions in season accommodation facilities at the time May 1st to June 30th and from September 1st to 30th, in some cases until October 30th; organizing of natural history bases at the time from November 1st to December 30th and from April 1st to May 1st; and organizing of the working people and youth winter stays in permanent accommodation facilities during the period of a permanent snow cover. It is possible to mitigate the extremity of season visiting of Orava Dam region (and lake regions on the whole) by the enlargement of these recreation forms.

An obstacle to their development is lack of initiative on the one hand, a weak material-technical basis for winter tourist traffic in the region of Oravska Magura, and absence of a coordinative organ of a general character on the other. This organ could help to overcome isolating ownership relations and develop other tourist traffic forms, and prolong the utilization of material-technical basis, which has been already completed, and thus turn this territory into an all-the-year-round lively region of tourist traffic. In this day, weekend tourist traffic from Zilina, Turiec and Liptov hollows may be activized and directed here; material-technical basis could be further developed and the effectiveness of completed tourist traffic facilities, as well as the expediency of expended investments could be increased. In this way it would be possible to contribute to man-power rehabilitation and the life style change of inhabitans in gravitating areas.

In this time the major Orava Dam gravitating zone is in particular Povazie, Hornad valley, and the adjacent region of

east and north Morava, a subsidiary one being other Slovakian
regions, a west part of Czech Socialist Republic (with more
visitors from Prague and bigger Czech towns) is of secondary
importance. The share of adjacent Orava region on Orava Dam
tourist traffic could be increased by tourist traffic activa-
tion and season prolongation and it would gradually spread
through the exchange or even sale of services to other districts of Slovak Socialist Republic and especially to industrial parts of Czech Socialist Republic. Change in the
structure of visitors would be dependent upon tendencies applied by the Orava Dam tourist traffic coordinating organ.

x   x   x

Finally it can be said that the disproportion problem
between possibilities of utilization lake districts facilities and their attendance (number of visitors) is often caused
by an unsatisfactory exploitation of social facilities in organizing stays in the very lake area and that of potential
conditions of adjacent regions. This problem can be solved.

Slater, P.B.
(USA, Morgantown, West Virginia)

HIERARCHICAL INTERNAL MIGRATION REGIONS
OF ITALY

A 1955-1970 internal migration table for 18 Italian regions is adjusted to have equal row and column sums. A hierarchical clustering algorithm is then applied to the doubly
standardized table obtained. This two-stage procedure has been
applied to internal migration tables for several nations.

Two sets of analyses are conducted here. In one (Fig. 1),
intraregional flows are taken into account. In the other
(Fig. 2), they are ignored. The most unusual result obtained
is the non-contiguous grouping of southern and northern
areas, before they join with central ones.

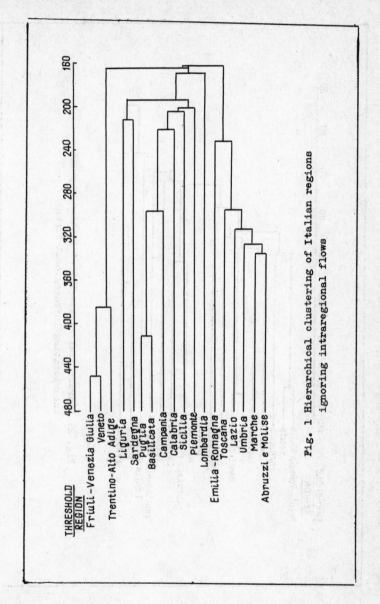

Fig. 1 Hierarchical clustering of Italian regions ignoring intraregional flows

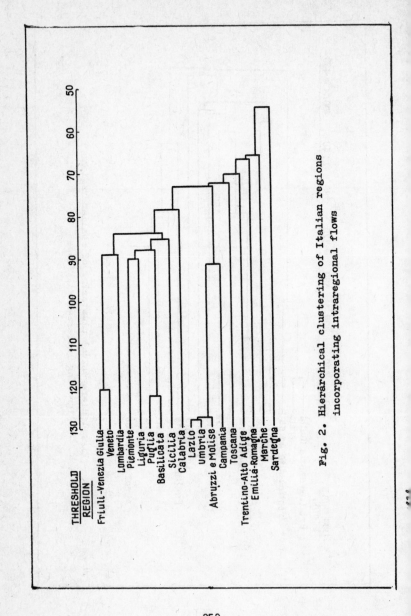

Fig. 2. Hierarchical clustering of Italian regions
incorporating intraregional flows

Sprincova S.
(ČSSR, Olomouc)

GEOGRAPHICAL ASPECTS OF SECOND-HOME LIVING

In developed countries massive migration of urban people into nature has been gaining more and more ground as part of rural/urban interaction. In ČSSR for instance, it concerns two thirds of citizens in their productive age. This phenomenon is related to advancing urbanization. In the past, man lived and worked in the country, now he comes here to spend his leisure to recover his physical and mental strength. Flows of recreationists making for the country are now becoming a sort of antistream of the permanent outmigration of rural people into towns.

One of the very pronounced forms this migration takes is second-home living. Its social, economic, and geographical aspects are now being more and more subjected to study and research. Let us only mention such authors as B.Barbier, F.Cribier, J.Maier, K.Ruppert, R.I.Wolfe, and others. The most frequent object of geographic research are the following aspects of second-home living: territorial relationships between permanent or first-home living and the places of non-urban recreation, evaluation of conditions necessary for the origin of second-home living, its demands for facilities, periodicity of the evoked migrations, associated activities, its effects upon environment and on social and economic structure of residential areas where it takes place. Relying on my research I should like to point to some of these aspects in the present communication.

ČSSR belongs to the countries of a relatively high amount of second homes, which holds good particularly for the Czech Socialist Republic. In 1972 the latter had 156,402 second homes with about 625,000 beds. This means 63 inhabitants per one home, or 16 per bed. The deciding condition for the choice of second-home site is, apart from distance and accessibility, the possibility of acquiring one's own second home. They are (like in Sweden, France, and many other countries) either houses formerly serving other purposes (farm houses and their premises, gamekeepers lodges, inns, fire brigade houses, etc.) within the villages suitable for recreation purposes. In ČSR there are 33,000 such properties estimated for further utilization of second-homing. Construction of new second homes is now planned and their distribution regulated by village planning respecting the protection and shaping of the landscape. It was only at the beginning of the whole process that those interested in second homes were able to choose the optimum terrain, climate and position for their recreation. Now being under the pressure of constant stricter regulations they are forced to cut down their demands or choose more and more distant places.

Second-home living has its specific requirements on infrastructure and facilities. The rising demands and changing life style is associated with greater call for the mains, sewage, quality of roads, garage or parking places. On the contrary, services and commercial centres are much less needed than in other forms of recreation and they may even be quite dispensable. This particularly holds good for weekend zones near towns, because second homers get their supplies in town and take them down in their own cars.

Second homing gives rise to very constant territorial association between first-home and second-home recreational ways of living. This relation is expressed in migrations of weekly or seasonal rhythm repeating with the regularity of the tide. These migrations make the number of present people in the respective villages rise on weekends and holiday to reach the limits given by the capacity of accommodation. It starts on Friday evening, culminates on Saturday afternoon, to fall down rapidly to the normal of local residents settlement on Sunday evening. The whole year trend of second-home living shows several peaks, namely during spring holidays, Easter, summer vacations, and Christmas holidays. Minimum values are found in November and December (till Christmas). Although realized in the same or analogous places as sporting accommodation (hunting, fishing, camping, in the past tramping), it is of a qualitatively different character. It is real living accompanied by activities otherwise typical for first-home living. But it stimulates its specific activity too (repairing the house or working around it), allowing at the same time other, especially recreational activity.

Second-home living is a factor influencing landscape environment and the social and economic structure of the respective villages. This is most remarkably reflected in the appearance of the landscape by building new dwellings which in most cases disturb the character of the countryside (e.g. new houses at the edge of the wood, on the banks of water areas, and on other exposed places). On the other hand, the use of empty houses inside the villages for second-home living purposes can be regarded as both economically (preservation and maintenance of housing) and aesthetically positive. Second-home living is usually associated with a number of other negative aspects (rubbish heaps, destruction of woods and meadows, rise in noise level, rousing animals, planting the gardens with plants not belonging to local plant community, etc.). This results in socially undesirable impairment of environment due to recreation, thus lowering its recreational value. We are now faced with the problem of protecting environment against recreation for recreation sake. Associated with this is the problem of capacity for second-home living. On the grounds of a field research in individual districts the bearable capacity of recreational living in CSR was estimated at 231,000 dwellings, only 75,000 more than the number reached in 1972. Considering that the number of second homes grows very quickly (e.g. from 1967 to 1972 it rose by 46,000), the limit will obviously be reached very soon.

If second-home living has negative influence on environment in most cases, its effect on the social and economic structure of respective settlements can be regarded as mostly positive. In consequence of industrialization and growth of

agricultural mass production and concentration of inhabitants
in towns, many villages founded in the times of feudal way of
production lose now their former function of centres of agricultural production and settlement. For those that have the
necessary conditions, second-home living becomes a new, socially important function. It is associated with a number of secondary phenomena. One of them is the replacement of the formerly rural residents by a new kind, the so-called "weekenders
and/or vacationists" spending here regularly their leisure
time, sometimes up to a third of all the days of the year.
These inhabitants often acquire a special relation to the village of their recreation stay, which becomes their second home,
and they themselves start showing interest in some aspects of
its prosperity (roads, services, shops, appearance, etc.). On
the other hand, they are not so much interested in the development of other forms of recreation in the place (such as mass
excursions of holiday makers, skiers, etc.).

In some countries the limit of second-home capacity will
be reached very soon and the demand will by far not be satis-
fied. That is why other forms must be considered, because the
problem cannot be solved only by forbidding the construction
of the further purpose build second homes. Structural geographical analysis of second-home living will no doubt become one
of the conditions for the perspective solution of the problem.

Stigliano M. - Moufarrej S.
(Zaire, Lubumbashi)

CARTES DE LA DISTRIBUTION ET DE LA DENSITE DE LA POPULATION
DU SHABA

Plusieurs territoires africains expérimentent ces derniers temps un remarquable changement socio-économique. Notamment, dans les pays dont les matières premières constituent
les ressources principales, les districts urbains nés à l'é-
poque coloniale sont devenus les pôles d'attraction d'une
masse croissante de ruraux. D'autre part la pauvreté des

256

structures du secteur secondaire fait de sorte qu'encore
aujourd'hui une grande portion de la population soit liée
au milieu rural, dont il est intéressant de connaître les
caractéristiques de quantité et de distribution des effectifs. Cela justifie ce travail qui consiste dans la construction et interprétation des cartes de distribution et de densité de la population africaine du Shaba. Il faut considérer que les données sont souvent incomplètes faute de documentation et à cause de la mauvaise organisation des liens
entre les centres périphériques et le chef-lieu, accrue par
un réseau routier en très mauvais état.

L'enquête est basée sur le recensement de 1970 et sur
les rapports des chefs de zones de 1971 et 72. Ces sources
montrent que 36,4% de la population habitent les territoires
urbains où ils trouvent des conditions de vie meilleures
liées à la possibilité d'obtenir un salaire.

L'évolution de la population des centres urbains dénon-
ce un accroissement moyen annuel d'environ 13%, contre l'accroissement moyen de la région d'environ 8%. Ceci confirme le
fait que l'exode rural présente des proportions remarquables
et crée ainsi toute une série de problèmes pour les centres
urbains vers lesquels ces effectifs se dirigent. Toutefois,
étant donné que ce phénomène est limité à 1/3 de la population régionale, les cartes conservent leur valeur puisqu'elles
se réfèrent au milieu rural où vivent encore 2/3 de la population africaine du Shaba.

En ce qui concerne la représentation cartographique, on
a d'abord construit la carte de distribution par points et
sphères, à laquelle on a superposé une maille de carrés de
5 mm de côté, équivalente à une superficie de 225 km^2. La
maille ne comprend pas les zones urbaines, les lacs et les
zones marécageuses. Les carrés qui ne contenaient aucun
point ont été éliminés mais comptés dans le calcul de la superficie totale. On a ainsi obtenu 1985 carrés qui représentent une superficie rurale du Shaba d'environ 446.00 km^2. La
population rurale, au nombre de 1.814.591 habitants, se dis-
tribue donc sur le territoire selon une densité moyenne de
4 hab./km^2.

Un examen plus détaillé de la carte montre que:

- plus de 60% du territoire est virtuellement inhabité;

- 1/4 du territoire contient 40% de la population rurale suivant une densité moyenne de 6 hab./km^2;

- 7% de la superficie est occupée par 35% de la population suivant une densité moyenne de 19 - 20 hab./cm^2;

- 25% de la population se concentrent sur moins de 3% du territoire suivant une densité de 30 à 70 hab./km^2.

Il apparaît donc qu'à un modèle de peuplement dispersé s'associent des régions densement peuplées, dont les limites peuvent être tracées grâce à la méthode adoptée pour la construction de la carte des densités. En regroupant les carrés vides contigus on identifie les zones inhabitées, dont la superficie est de 52.000 km^2 (12% du territoire): il s'agit des zones des parcs nationaux, des zones marécageuses et d'escarpement. Le regroupement des carrés à densité égale permet de délimiter trois régions de peuplement, dont la plus importante correspond au sillon du Lualaba, les deux autres étant le long du lac Tanganyika (pêche) et le long de l'axe ferroviaire Likasi - Lobito (agriculture).

En guise de conclusion, l'examen des cartes nous permet de définir deux modèles de peuplement qui ont intéressé successivement le Shaba: le premier, témoigné par les trois régions à peuplement dense, est d'époque précoloniale; on lui a superposé un modèle colonial obéissant à des exigences totalement différentes, qui apparaît étranger à la réalité du territoire et toutefois, comme on disait au début, conditionne tout le Shaba en tant que pôle d'attraction pour une large masse de ruraux.

Federico Sulroca
Calixto Machado
(Santa Clara, Cuba)

AN URBAN DEVELOPMENT STUDY BASED ON THE STUDY CASE OF THE CITY OF SANTA CLARA

The analysis of the natural environment bordering an urban nucleus represents a basic premise for the future evaluation and planning of the perspective development.

In the urban network studies of the Cienfuegos region, we applied a methodology which was further developed and used for the city of Santa Clara, capital of the province Las Villas, where it proved to offer satisfactory results.

The principle aims of this study, as well as the results obtained, are summarised in the following points:

1. Evaluation of the environmental building qualities.

In which the main characteristics of the territory according to construction possibilities are synthesized, considering the following aspects:

(a) Geological, geomorphological and geological engineering conditions;

(b) Edapho-pedological conditions;

(c) Geo-hydrological conditions.

For this purpose, we traced an area with a radius of 10 km around the city Santa Clara. Using this for our study, we made: a Geological Scheme, a Dismemberment Vertical Relief Chart, a chart on the density of relief outlines, diagrams of the zonal fracturation system, a Slope Map, a Scheme of Relative Intensity of Erosion, profiles, Neotectonic Charts, soil maps, a chart on soil resistance for construction. At the same time, making up the respective evaluation tables referring to constructive aspects. In addition we considered the analysis of the water horizon depth, the superficial and subterranean seepage system, the zones affected by floods, the territory's agricultural value, as well as possible mineral and forestral resources, etc.

With all these elements, we made a summary chart, evaluating and zoning the territory according to different constructive limits.

The results we obtained, permitted us to determine a natural threshold towards the southern and eastern limits of the city, represented by the hilly grounds of the "Gran Intrusion Central" (Great Central Intrusion), we also established a favourable zone for constructive developments, coinciding very much with the present urban area and its industrial belt, a zone towards the west with partial borders for major-constructions allowing at the same time for minor constructions, a band towards the south and the east with strong limits in its relief and geological engineering conditions, which constitute an actual threshold in accordance to our construction techniques, and a zone towards the North, aftward soil resistance.

2. Evaluation of the Environment and of the Sanitary and Hygienic Conditions

The aim of this point, is to determine the climatic characteristics and contamination of the environment, as a function of urban ecology, and to establish the sanitary policy in perspective urban growth.

To this purpose we processed informations on the different climatic variables, with more than 10 years observation, summarizing the results in a table, and made a climogram including the relation between temperature and rainfall and an evaluating table of the climatic characteristics for periods and sub-periods according to urban conditions.

We also analysed the industrial problems and their relation to environmental pollution, the problems concerning water pollution, waste elimination, etc.

As a result of this analysis, we reached the conclusion, that the climatic conditions of Santa Clara are by no means favourable for urban growth. Therefore, a development of green belts ought to be encouraged, as well as perspective forests surrounding, the industrial zones. In addition to this, we reached the conclusion, that spaces between the buildings ought to be increased, especially in the housing and development zones, that industries noxious to the environment be prohibited, that the construction of a new sewer system should be planned, that the streams crossing the city be canalized and that the construction of recreative centres (covered with a roof) be encouraged.

3. Urban Hydraulic Balance

The same was done, bearing in mind the actual and perspective situation, the regional water supply system, and the hydraulic works programme. The results proved, that the perspective urban development is guaranteed, although this implies a complex and costly programme of hydraulic works, because the territorial potential is poor, especially that of the subterranean water resources. Therefore we do not recommend the acceptance of high indices of water consumption.

In summarizing these three factors, we elaborated the national potential of Santa Clara in view of the urban development, which according to the directives of the development which was planned, allows us to elaborate proposals for the Territorial Organization.

Based on all the above mentioned facts, we understand, that the urban growth of this city is expressly linked with its natural conditions, which are not very favourable for these purposes. Therefore, we should evaluate the expansion towards the west, other infrastructural and economic parameters, as well as study in detail, all the other nuclei of the regional urban network, in order to conform a complex evaluation of the territory which allows us to trace a development policy.

We also recommend the use of morphometric methods for the analysis of the natural conditions, according to the experiences we obtained.

Toušek V.

(ČSSR, Brno)

THE ECONOMIC STRUCTURE OF COMMUNITIES AND ITS CHANGES DURING 1961 - 1970

The Czechoslovak pre-war settlement geography could not pay attention to the economic structure of settlements owing to the lack of statistical data and focused mainly on the study of their physiognomy. But this gives an imperfect idea of the population structure on which it depends directly. The economic structure of the settlements and its changes show distinctly the development of the productive power of human society, the character of the system and its social-economic organization.

The data concerning the economic structure of all Czechoslovak communities are involved in the census 1950 and, in greater detail, in those of 1961 and 1970. Thus, conditions were created for the analysis of the employment structure of communities which is one of the important characteristics of the function of the settlement. The economic structure is a frequently applied criterion in the geographical classification of settlements in works dealing with the economic-geographical regionalization of Czechoslovakia. The typization of the communities is based either on the structure of the economically active population working in the community or living in the community and their share in the most important economic branches. In the paper the latter method has been applied on the example of the South Moravian Region which represents suitably the spatial distribution of the economic types of communities in Czechoslovakia.

Table 1

Structure of the economically active population of the South Moravian Region according to sector (1961 and 1970)

Sector	1961	1970	Difference
primary	26.8%	18.4%	- 8.4%
secondary	49.2%	50.0%	+ 1.5%
tertiary	24.0%	30.9%	+ 6.9%

Table 2
Economic types of the communities of the South Moravian
Region

Type	Number of communities	
	1961	1970
1. (a) agricultural (A)	208 (13.1%)	87 (5.5%)
(b) agric-industrial (AI)	671 (42.2%)	468 (29.4%)
(c) agric-services (AS)	13 (0.8%)	15 (0.9%)
(d) mixed with prevailing Ist sector (Ma)	58 (3.6%)	128 (8.1%)
2. (a) industrial (I)	41 (2.6%)	80 (5.0%)
(b) industrial-agricultural (IA)	420 (26.4%)	339 (21.3%)
(c) industrial-services (IS)	75 (4.7%)	198 (12.4%)
(d) mixed with prevailing IInd sector (Mi)	96 (6.0%)	255 (16.0%)
3. (a) services-industrial (SI)	5 (0.3%)	9 (0.6%)
(b) mixed with prevailing IIIrd sector (Ms)	4 (0.3%)	12 (0.8%)
Altogether	1591 (100%)	1591 (100%)

The limit value for the establishment of the economic
type of the communities is the share of 20% of economically
active persons working in one of the basic economic sector.
For the purpose of comparison the number of communities of
1961 was converted into that of 1970.

From the data in the tables it follows that the changes
in the economic structure owing to industrialization, bring-
ing the collectivization of land stock to the end, the dynamics of the technical equipment of work in agriculture, development of transport (1/3 of working people commuting daily
to work), migration to larger communities and mainly the development of the tertiary sector manifested themselves to a
considerable extent by the change of the economic types of
communities.

Following conclusions can be drawn from the comparison of the economic types of communities:

(1) Complex view. A change of the economic type took place in the case of 812 communities (51%). The number of the communities of types A, AI and IA decreases considerably. Most communities (149) changed type AI in IA, 112 communities type A in AI and 108 communities type IA in Mi. The highest increase appears with communities of types Mi and IS.

(2) The economic type and size of the community. Types AI and A prevail with communities up to 500 inhabitants. In the group of communities with 500 to 999 inhabitants type IA prevails in the group of communities with 1000 up to 1999 inhabitants types Mi and IS predominate though in 1961 a prevalence of communities of IA type was recorded. In the size group with more than 2000 inhabitants most communities are of IS type.

(3) Spatial distribution of economic types. In the districts of Zdar nad Sazavou, Znojmo, Trebic, Jihlava, and Breclav there is a considerable number of collective farmers which results in the majority of communities of types AI and A. In other districts communities of IA type prevail. The influence of the town Brno manifests itself most in the district of Brno-province (38% of communities are of IS type).

(4) Prospect. In the following decade when the continuation of the present increase of people working in the IIIrd sector and a planned decrease of people working in agriculture are expected the present-day trend in the change of the economic structure i.e. the decrease of the number of communities of AI type and the increase of communities of IS and Mi types can be supposed to keep preserved.

VIII. REGIONAL GEOGRAPHY

Burton I.
(Canada, Toronto)

A ROLE FOR GEOGRAPHICAL STUDIES OF ENVIRONMENTAL
PERCEPTION IN INTERNATIONAL DEVELOPMENT

An opportunity exists for geographers to build on the recent development of concepts, methodology, and techniques for
the study of environmental perception in contributing to the
understanding of some of the more intractable problems of modernization faced by nations of the third world. Environmental
perception is now recognized at the international level as a
concept that bears in an important way upon the choices individuals and groups make in dealing with an uncertain environment, and upon the differences that are to be found between
the approaches adopted by outside experts or interviewers and
those who exercise the prerogatives of choice. This paper
describes the importance of perception variables in relation
to two specific areas of development - the management of natural hazards, and the improvement of rural water supplies -
and then outlines the efforts being made within the UNESCO Man
and the Biosphere Programme to draw upon such lines of investigation.

Perception of Natural Hazards

Extensive research on human adjustment to natural hazards
has shown that reliance on the use of technology for the control of environmental processes can exacerbate the problems
that it was intended to alleviate. (White, editor 1975. Bur-
ton, Kates and White. Forthcoming). Construction of barriers

and cross-dams on the outer "chars" or islands in the Ganges-Brahmaputra delta of Bangladesh, for example, has served to encourage further encroachment into such hazardous locations with fearful consequences in terms of lives lost in tropical cyclone floods. (Islam, 1970) Similarly the use of irrigation and new agricultural methods has served to increase the vulnerability of parts of East Africa to severe drought.

Better knowledge of the process of perception of such natural hazards and the choice of adjustments to them can help in the creation of a pattern of development which curtails the further growth of damage and reduces loss of life.

Rural Water Supplies

Efforts to improve the standards of quality, accessibility and reliability of rural water supplies in developing countries have encountered considerable difficulties. The magnitude of the problem is extremely large (over 1,000 million not adequately supplied in 1970) and the cost of improvements for such a population is extremely high in relation to the present allocation of resources. The 1970 level of expenditure for construction costs in about 90 developing countries was estimated at $138 million. This would need to be more than doubled to $280 million annually if the World Health Organization's goal of serving 200 million additional people for each year during the Second U.N. Development Decade is to be achieved. The prospects of achieving such a goal are slim, and even if this were possible those remaining without safe and adequate water supplies would still be a greater number in 1980 than in 1970 due to population increase.

There are divergent reasons for slow progress, but a significant proportion of them have to do with the differences in perception between village populations (the users) and the national authorities responsible for rural water supply programmes (the delivery system). Preliminary studies indicate that users perceive alternative sources of water in a different light than the water supply engineers or public health officials. Commonly the users are guided in their choice by a much wider range of social and cultural variables than those which habitually enter into engineering decisions. Much the same appears to be true of the choice of technology or the design and level of the actual improvements to be made. When decisions are made by the delivery-system that fail to take local user perceptions into account it is hardly surprising that improvements fail for want of adequate maintenance or are sometimes simply abandoned and not used even when in working order. The introduction of perception studies on a systematic basis could help to provide new information for the delivery-system which would then have an opportunity to respond more effectively to local needs. To achieve this will require the creation of some successful models or examples of how perception studies can be made to serve in this manner, plus a substantial input of political will or motivation at the national level. Efforts are now underway through the International Development Research Centre (Ottawa) and the inter-agency Ad-Hoc Working Group for Rural Potable Water Supply and Sanitation to encourage and support the necessary studies on the social and managerial aspects of rural water supply, beginning it is hoped with some studies in Africa. If such research can demonstrate that work on the differential perceptions of users and the delivery system pays off in terms of the rate at which improvements can be made and in the long-term viability of such improvements then the political will should not be hard to find. In the absence of such results the rural water supply field will continue to look unattractive as an area for the investment of scarce financial and skilled manpower resources.

A Paradigm for Perception Studies

As the Programme of UNESCO for studies of Man and the Biosphere took shape it was recognized that environmental perception was a crucial element in achieving the general objective of the programme.

Subsequent to a meeting of experts on Project 13 (UNESCO 1973), geographers and social scientists with knowledge in the field of perception studies have contributed to project development and formulation in relation to several MAB Pro-

jects. To further this work and to assist in the incorporation
of perception studies in MAB Projects at the field level a
handbook is being prepared entitled <u>Guidelines for Field
Studies in Environmental Perception</u>. The <u>Guidelines</u> will fulfil two main functions:

(a) provide a rationale and description of the field of
environmental perception in the context of man-biosphere relations and ecosystem management, and

(b) suggest alternative research methods for field investigations of environmental perception.

As the <u>Guidelines</u> become available it is hoped that they
will serve as a tool for geographers to use in developing
perception studies and in contributing to the understanding
of some of the more intractable problems of modernization
faced by nations of the third world.

Eringis K., Bumblauskis T., Pakalnis R.
(USSR, Vilnius)

PROBLEMS AND PRINCIPLES OF ENSURING THE QUALITY
AND STABILITY OF LANDSCAPE WHEN CONSTRUCTING
FLOW-REGULATING PONDS (UNDER THE EXAMPLE OF
LITHUANIA CONDITIONS)

The ancient primitive system of flow-regulation by means
of small ponds and reservoirs of water-mills in Lithuania
lost its significance in the period of 1953-1975. Now, in relatively rich with lakes Lithuania in definite localities
quite often encounter the limits of water resources. In the
light of this, lakes are used more intensevely and 440 new
artificial ponds of different size and number of water-supply
systems of cities have been established. In future in Lithuania it is possible technically to increase the useful volume
of the flow-regulating reservoirs. However, constructing and
utilizing flow-regulating ponds the spatial heterogeneity of
geographical conditions shows up. The ecological situation

stands complicated in the valleys especially. Moreover, the interference of Man in to the motion of natural processes when utilizing water resources violates natural relations more and more, and increases areas undesirable in ecological and aesthetical aspects.

Under above mentioned conditions and situations the new phenomena in landscape come into effect. Firstly, lately the areas build on with different urban, industrial and agricultural units, roads have been increasing rapidly and the areas of arable land have been reducing. Thus, the protection of land for agriculture when regulating flowing acquires particular significance. Secondly, the production and utilization of materials, which make ponds polluted increases. Besides, artificial ponds, in comparison with lakes, in the landscapes life have a short-term character. After some 100-200 years, at 0.5-1% silting up they will stop to be useful water units. It should be noted that hydroenergetic reservoirs such as Kauno-Marios and Antaliepte in Lithuania (on the rivers Nemunas and Shventoji) gradually approximately during a decade lost its energetic signifance, because of insufficient providing them with waters resources. In the third, for the first time in such a scale in our republic science and practice have come across a new water environment and medium the significance and properties of which change more rapid than one can realize and forsee its future. This environment becomes more and more important as the unit of recreation with its complex utilization. And this problem can't be solved there by one field of knowledge. With indispensable width it may be solved by interdisciplinary approach only. In the fourth, under the conditions of moderate damp climate technogenesis begets not only and not so much the problems of water resources and hydroenergetics, as the problem of protection of the most valuable natural units and resources, the monuments of Nature of the high-qualified and stable landscape, pure water and its space, the territories of recreation and, on the whole, the environment inhabited by Man.

High effectiveness of agricultural utilization of land in Lithuania, increasing of deficit of fertile soil, and also difficulties coming into being because of maintaining normal biochemical water regime, make the projects of construction and utilization of the large wateraccumulating flow-regulating reservoirs doubtful. Things are different when constructing relatively middle size or small artificial ponds and lakes with complex utilization, which practically are isolated from large river systems and more controlled. We have even historical experience in construction and utilization of small artificial ponds. Now, after all, there are worked out some scientifically-founded principles and methods of purposeful formation of landscape in the sphere of influence of flow-regulating ponds and the construction of model-experimental ponds is carried out (Пакальнис, 1971; Eringis and otr., 1971; Бумблаускис, 1975, Экология и эстетика ландшафта, 1975). It is ascertained, that after the corresponding ecological and geographical investigation it is necessary to carry out protecting arrangements in each artificially-created pond on the riverine or lake-riverine system. From our joint investigations and the constructive revisions of Lithuanian Research Institute of Hydrotechnics and Land-reclamation follows the possibility of the adaptation of differentiative approach while forming the environment of highqualified pond depending on natural conditions and antropogenic demands. Changing the purposeful geographical environment by means of construction and utilization of flow-regulating ponds and by their surroundings it is important to maintain the following ecological principles: a) the optimization of ponds-leveled regime, especially in the period of vegetation; b) diminuation to the minimum of the area of shallow sections of ponds, and the section of the hollow of pond, which comes forward from under the water when the pond functions. It may be achieved by the stabilization of high water-level in separate parts of ponds, using for all that the dams coming forward from under the water and additional simple weirs. Besides, the small ponds-satellites of the main pond, and polders of different type should be laid in the ravines and valleys of tributaries (ponds are laid with the stable or slight changing water-

level; polders are laid of different size using the sluice-
regulaters, pumping stations or derivating canals); c) the
utilization of small ponds-satellites of the main pond for
accumulation of migrating materials which increase the eutro-
phication of ponds; d) preservation of humus layer of soil,
preventing peat reserves and sapropel from flooding, depending
of the important part of shallow sections; e) ensuring of
preservation of unique landscape units; f) instillation of
conditionally exclusive cycles of water utilization on order
to ensure preservation of selfregulation of water ecosystems;
g) in water-collecting basin the economic activities must
become farsighted without detriment to a pond.

Complex interdisciplinary investigations when studying
the problem of flow-regulating water systems of separate areas
is regional planning basis for the definition of qualified
and lasting ways of mastering and economic utilization of
geographical landscapes. The successful solution of this problem and putting into practice the principles and measures of
protection of ponds are practicable only by the wide interdisciplinary approach. One of the main tasks of constructive
geography while regulating flow and preserving the natural
environment is to find such a solution of complicating technical tasks that, when translating them into reality the Nature and technosphere and human culture might develop.

Gaspar J.
(Portugal, Lisbon)

THE GEOGRAPHY OF PORTUGUESE ELECTIONS
(25/4/75)

There were no studies in electoral geography during the
dictatorship that prevailed in Portugal from 1926 to 1974.

We made a research on the geography of the elections for
the Constituent Assembly that took place on the 25th of April
1975. This reasearch has naturally much to do with sociology.
In this paper we present in summary some of the conclusions
we reached.

1. The parties. 12 parties showed themselves as candidates although only five competed in all the provinces (18 in the continental part of Portugal). This fact does not make certain methods of statistical analysis very easy.

An analysis of the social and professional structure of the lists of candidates in each party, connected with the hierarchic position of the several professions on those lists and at the same time with the professional composition of the lists in each province, permitted a first arrangement of the parties from the right to the left wing. Greater participation of the workers took place in the left wing parties, as opposed to the upper and middle staff in the right wing parties. With only one exception, of least significance, the arrangement of the parties obtained through the method we used, agreed with the generally accepted classification, i.e. the one suggested by the means of social communication. We do not distinguish by means of this analysis very neat groups of parties, only a steady opposition of the right to the left wing, where the Communist Party has a relevant place.

Another kind of analysis that lead to the arrangement of the parties in terms of right versus left wing, was the study of the contents of the electoral propaganda through the daily newspapers. The contents was evaluated on the basis of the occurrence of 100 key-words, values which have subsequently been processed through the factorial analysis.

In this approach we also obtained an arrangement of the parties from the right to the left wing, which coincided almost entirely with the one mentioned above. While in the first instance there seems to be a steadiness, in this one we observe a grouping process. The linkage tree obtained shows in the first place two main groups. The first one includes the parties from the right wing to the Communist party, arranged according to the commonly accepted order; thus the party furthest to the right (CDS) joins the monarchists, while the social democrats (PPD) join the socialists (PS), the next one to follow this block of four is MDP (a unity party without neat political and historial steadiness) and finally the Communist Party.

2. Aspects of the electorate. From a population of
8.5 million, excluding the emigrants and citizens of the co-
lonies, enroled in the electoral registers 6.2 million citizens (73%), which represents more than 3 times the number of
any preceding elections.

Every citizen over 18 in possession of his political
rights was able to enrol himself. The number of enrolments
in the electoral registers was high in every part of the
country, sometimes higher than the maximum expected. Meanwhile we noticed some discrepancies (dichotomies). So, as a
rule, the number of enrolments in the lists was higher in
the south than in the north, mainly in those areas where there
is a predominance of large estates and was higher in urban
rather than in rural areas.

The electors chose a fixed number of candidates in each
electoral area, each coinciding with the province. Each de-
puty represented 25,000 electors enrolled or a fraction of
more than 12.500. The amount of voters in the elections was
very high (91.7% of enrolments number).

The Geographical distribution of the voters

The figures obtained by the 12 parties were very different, 37.9% (PS), 26.4% (PPD) and 12.5% (PC) for the parties
with more votes and a total of 16.3% for the seven parties
with less votes. The votes not used represented 6.9% of the
total. This great dissimilarity in the total figures corresponded to very strong regional and local dissymetries.

Meanwhile emerges a clear tendency: the one opposing
north to south. A net opposition between rural and urban areas
was evidenced, though it is not identical throughout the country and shows also a dichotomy north-south.

This tendency north-south corresponds to a more conservation north and a more progressive south. This dichotomy
however is not entirely true. The more progressive provinces
are the four in which there is a predominance of large estates, south of the Tagus River and the Lisbon industrial
belt. But the votes in the Algarve, the southern most district.
resemble more those of the central districts, which also follow the same tendency north-south.

273

The difference in electoral behaviour coincided approximately with a dissimilarity in the participation: the number of non-voters was higher in the north than in the south.

None of the parties had an identical behaviour from north to south. The biggest divisions occurred in the Communist party ($\overline{M}$ = 12.5%, Sd = 12.5) and in the two parties which had tendencies most oriented to the right - CDS ($\overline{M}$ = 7.6%, Sd = = 5.7) and PPD ($\overline{M}$ = 26.4%; Sd = 13.9). We excluded the small monarchist party, PPM, which did not compete in all the provinces. The socialist Party had regionally the most balanced number of votes ($\overline{M}$ = 37.9%; Sd = 8.7) with a minimum of 21.3% in the province of Vissu and a maximum of 52.4% in the Portalegre province. While the Communist Party had a minimum in the north (2.3%, also in the Viseu province) and a maximum in the south (39% in the Beja province), the socialist party had its lowest figures in the north and the highest in the centre districts as well as in the Algarve. The PPD had its highest figures in the north (42.9% in Aveiro) and the lowest in the south (5.2% in Beja).

The parties which lie to the left of the communist Party, though stronger in the south had great differences either at the province or at the commune level, obtaining by their addition together the highest figures in the centre province of Castelo Branco (6.6%). Some parties of the extreme left wing have even obtained their highest numbers of votes in the communes or even provinces of the north.

Considering only the five parties that competed in all the electoral areas those whose votes are more similarly distributed in space are the two parties which lie furthest right, CDS and PPD, with a correlation coefficient (R) of + + 0.76.

On the contrary the most dissimilar are the P.P.D. and the P.C. (R = -0.88), being the values of R for the pairs CDS/PS (-0.73) and CDS/PC (-0.69), still very high.

The correlation between the results of the elections and the social and economical standards gives usually extremely low values.

274

In fact, though the big dichotomy (discrepancy) in the votes lies in the opposition north/south, the Portuguese dichotomy concerning development lies between the coastal fringe that ranges from the southern part of Lisbon to the northern part of Porto, and the rest of the country both the north and the south.

Although the voting in the northern coastal fringe is less conservative than in the inner regions, the same is not true of the centre and south.

Meanwhile, the correlation between the votes and the average size of the rural estates is sometimes rather high. Thus for the number of votes obtained by the Communist Party, the value of R is +0.93, for PPD it is - 0.77 and for CDS it is - 0.56.

While in the areas of large estates most people voted for the Communist Party, in the areas with predominance of small estates people preferred the more conservative parties. The correlation between the votes obtained by the Socialist Party and the average size of the rural estates produces a rather low though still positive value (R = +0.26).

In vue of the restraints of this summary, we searched for a map that could best represent the vote distribution of these elections. We chose to distribute the number of votes in the communes in three significant groups:

(a) The total sum of the parties that lie to the right of the Socialist Party (CDS, PPM, PPD).

(b) The Socialist Party.

(c) The total sum of the parties that lie to the left of the Socialist Party - the remaining eight.

The right wing dominates in the north excepting in the metropolitan area of Porto and its domination is interrupted in the coast by a stain of the PS whose focus is Coimbra. The Socialist Party dominates in the centre, in Algarve, in the metropolitan area of Porto, in Coimbra and in small industrial areas of the north as S. Joao da Madeira and Gouveia. The northernmost commune gained by the PS, Caminha, has an old tradition in the opposition even since the time of the fascism. The PS wins also in the northern part of the metro-

politan area of Lisbon - where there is a predominance of the
tertiary activities - and in the northern part of Alentejo
(large estates). The left wing dominates in the Alentejo and
in the industrial periphery of Lisbon, especially south of
the Tagus River. An analysis at the parish level, however,
shows its superiority in almost the entire industrial sector
of the metropolitan area of Lisbon.

The northernmost commune still giving advantage to the
left wing, corresponds to an industrial agglomeration (mainly glass works) with an old Communist tradition. On the other
hand, the southernmost commune voting for the left (Vila
Real St. Antonio) corresponds to an urban agglomeration of
fishing industries.

Klimek K., Lomborintchen R., Starkel L., Sugar T.
(Poland, Cracow; Mongolia,
Ulan Bator)

RESEARCH PROBLEMS OF THE MONGOLIAN = POLISH GEOGRAPHICAL EXPEDITION TO THE KHANGAY MTS IN 1974

The Mongolian = Polish Physical Geographical Expedition
to the Khangay in 1974 was organized within the framework of
scientific cooperation between the Institute of Geography,
Polish Academy of Science and the Institute of Geography and
Geocryology, Academy of Science of the Mongolian Peoples Republic, as a first stage of two years expeditionary investigations. The main purpose of the expedition was to learn about
the resources of the geographical environment on the southern
slope of the Khangay from the viewpoint of possible advances
in the national economy, especially its animal hushan-dry.
Fifteen scientists representing various branches of the earth
sciences particiced in the expedition including geomorphologists, hydrologists, climatologists, botanist, pedalogist and
geodesist.

The main base of the expedition was established at the mouth of the valley of the Tsagan - Turutuin - gol, where it passes from the mountains into the foreland at about 2100 m. A sub = base of the expedition was founded in the southern part of the depression basin of Bayanenurin - hotner situated in the foreland of the Khangay at a height of 1960 m. The work of the expedition in the Tsagan - Turutuin - gol basin lasted from June 15 to July 31, 1974. The study area of the expedition included the basin of the Tsagan - Turutuin - gol, the surface of which is 2176 km^2. At the expedition base a climatological station and a hydrological station were established. Owing to the location of the base it was possible to perform studies also in the Khangay and on the nearby mountain foreland.

The mountainous character of the research area caused the participants to divide into two problem parties; one working within mountain slopes and other within valley and basin bottoms. In view of the large size of the research area and at the same time the necessity of satisfactorily detailed acquaintance with various elements of the geographical environment and the processes at work there, in both parties carried out parallel investigations of both general and detailed character.

The "slope" party under the guidance of L. Starkel first of all carried out detailed investigations of nearly all the elements of the geographical environment and relations between them in the Sant valley opening to the valley of the Tsagan - Turutuin - gol in the marginal (southern) portion of the Khangay. The investigations aimed at discovering the mutual relationship between relief, macroclimatic conditions, water circulation, soil and plant cover on north and south - fac̆ing slopes in various climatic = vegetation belts. Outside the Sant valley the participants of that group also performed general or comparative studies on slopes in other parts the Tsahan - Turutuin - gol basin.

The "valley" party led by K. Klimek carried out general and detailed studies the aim of which was to investigate geomorphological problems, climate, water conditions and certain

present – day geomorphological processes, mainly in the valley of the Tsagan – Turutuin – gol and in the valleys of its
major tributaries.

Both parties were at the same time facing practical
tasks. For the slope party such a task was to become acquainted with the types of geo- and ecosystems, their abundance,
and the degree of their degradation as the consequence of the
intensification of animal husbandry.

The principal task of the valley group was, to determine
water resources and possibility of water supply to animal
husbandry as well as to villages and processing industry.

The expedition will continue the investigations in the
summer of 1975.

Kluge K.
(GDR, Neubrandenburg)

DEVELOPMENT OF AGRICULTURE AND ITS EFFECTS ON
SPATIAL STRUCTURE IN RURAL TERRITORIES

In all countries of the community of socialist states
development of agriculture is characterized by continuous
growth.

The Eighth Congress of the Socialist Unity Party of Germany had oriented agriculture towards further socialist intensification. That means a transition to industrial forms of
production by way of cooperation, what in fact is a revolu-
tionary process. In the agricultural territories of the GDR
have been constructed many plants for large-scale stock-farming, crop production is highly mechanized, relations in production are developing into the form of so called "cooperative divisions of crop production", by joining LPG (agricultural cooperatives) and VEG (state farms).

It is obvious that this new phase, going on by way of
cooperation and being characterized by concentration and specialization as well as by intensification of interrelations
between agriculture and food processing, has intrinsic ef-

278

fects on social development of the respective territories, requiring thus a certain specialization of agricultural areas.

With the aim of planning this development in its complexity, especially in the district of Neubrandenburg, have been elaborated many "Conceptions of Territorial Development", in which are coordinated all social and economic processes including those going on in settlement structure.

There can be noticed the development of the following areas with specific main functions, giving base to new possibilities of economic regionalization:

1. Areas dominated by enterprises for large-scale stock-farming including plant production which specializes in provisioning these farms. These areas have a greater number of agricultural labour (ca. 10 - 14 workers per 100 hectares of land in cultivation) and therefore of population, too.

2. Areas specialized in plant production without large-scale stock-farming, but using and modernizing available equipment for stock-raising; number of labour will remain constant at 8 - 12 persons per ha.

3. Areas specialized in plant production without large-scale stock-farming and with diminishing of yet existing traditional stock-farming; number of persons employed in agriculture declines to 4-6 per ha.

4. Agricultural-industrial complexes, developing in connection with larger towns. By capacities for forage production to be developed here, will be provided all the large-scale enterprises of stock-farming in the hinterlands of the respective towns. Viceversa products of these enterprises are processed in those towns for consumption in town and hinterland.

5. Recreational areas, with an agriculture being specialized in crop production in such a way, that undesirable effects on recreation will be avoided. Number of agricultural workers will decrease here, too.

Settlement network in the rural districts of the GDR does not include - as resulting from a low degree of industrialization - any larger agglomerations. Level of urbaniza-

279

tion is therefore relatively low, settlement network reflecting hierarchical structure.

Settlement pattern in the district of Neubrandenburg is characterized by a radial structure, the city of Neubrandenburg lying in the very centre, with direct connections to the other county towns around it. These county towns and the remaining small central places are in their turn located on other nodal points of this radial structured settlement network.

Small towns and rural central places are developing more and more as centres of agriculturally specialized areas and, resulting from this, as centres of "Gemeindeverbande" (a recently developing form of associated communities), which are constituting parallel to extending co-productive relations in agriculture. That means: starting from processes going on in agricultural production, all parts of social life in these associated communities will be developed complexly.

Within the types of areas mentioned, process of concentration will continue, the following main trends being recognizable:

- Enlargement of the functional basis of the rural centres

- Stabilization of functions in those villages, where administration for a community is located

- Decrease of functions and population in the other settlements, the smallest of them, that means in general those with most unfavourable working and living conditions, being on the way to extinction.

This development will further contribute to eliminating the differences between town and country.

Marková C.
(ČSSR, Prague)

THE CONSTRUCTION OF THE RANCHI HEAVY ENGINEERING
COMPLEX (INDIA) AND THE LABOUR FORCE STABILIZATION
PROBLEM

Not far from the largest Indian steel plant in Bokaro,
which is in the stage of completion with Soviet assistance,
two socialist countries, the USSR and CSSR, are building a
heavy engineering complex near the town of Ranchi in state
Bihar. The site of the object was chosen by Soviet experts,
who also built the largest part of the complex - the Heavy
Machine Building Plant with an annual capacity of 80,000 tons.
Czechoslovakia built the Foundry Forge and Heavy Machine
Tools Plants.

The construction took place in very complicated circumstances, in which usual obstacles were aggravated by a number
of specific regional geographic conditions, as well as the
still open question of suitable labour force. The matter of
concern is due not only to the low cultural level and high
natural population growth, but also to the fact that the construction of the large industrial plants incited an enormous
influx of people from the widest vicinity, often also from
distances of hundreds of kilometres (from Bihar, Orissa,
Madhya Pradesh, and former emigrants from present-day Bangladesh), peasants, craftsmen, the town poor, people very diverse in nationalities, languages, with deeply ingrained religious prejudices. They search for labour opportunities in the
new modern plants. In a short span of time they are expected
to shape into a uniform type of industrial employee capable
of participating in the complex production operations of
plants equipped with up-to-date technology. At the same time,
however, individual work groups incline to mutual alienation,
to the observance of caste customs and ancient traditions,
religious feasts, to the protracted celebration of family
events, and so on. It is considered a matter of course that

281

in spite of regular employment at the plant, these workers will return to their native villages for several weeks at harvest time to help with agricultural work. The workers' average absence rate is 16—22% of the overall number of working days. Feasts and holidays account for 15% of absence from work, agricultural activities, mainly the harvest 10%, accidents 22%, illness 27%, drunkenness 7%, transport difficulties 9%, hard or dangerous work 10%.

Work operations are also grossly obstructed by population migration. In the Ranchi region only 37% of the inhabitants are native. The immigrants (63%) usually do not stay long. Especially tribe members, Munda, Oraon, Kola and others, who represent 20% of the Heavy Engineering Complex workers, adapt themselves only with great difficulties to conditions of large plant production. After initial training many leave the plant because of inability to conform to the new style and regular pace of life. Of the overall number of industrial employees in Ranchi, not more than 45% had ever worked in industry before. The plants invest considerable sums into the training of employees, nevertheless it requires great effort to obtain stable labour. The productivity per head is not higher than 40%. Operations run on less than half the capacity.

There is a similar situation in other parts of India, too, although not in such sharp form as in Ranchi. It is becoming obvious that one of the preconditions of successful industrialization is the availability of sufficient suitable labour force. This can be achieved only by increasing investment on a national scale into man as such. Not only by improving nutrition, but also health care, schooling opportunities, and everything else that young people need. It is necessary to achieve a more stable balance between investment into industry and into people in the whole country. Only with a sufficient number of qualified workers and technical cadres will it be possible for the plants to avail themselves fully of their operation capacity and include more complex products into their processing programs without ensuing problems.

The example of Ranchi proves the unfavourable impact in some developing countries of the gap between the painstakingly elaborated theoretical principles of regional development and inadequate conditions for their practical application. The economic geographer should estimate in each concrete case the chances for the realization of theoretical decisions, especially in the case of construction of big-investment industrial units. Construction should be preceded by a regional study including a detailed analysis of the population and its development in a 100 km radius from the planned location. The research should concern men only, their professions, ages, birthplaces, accomplished levels of education, nationalities, religions, mother languages, knowledge of other languages. Leaning on a detailed analysis of gathered data, an experienced geographer familiar with local conditions can specify narrower regions with the largest availability of stable labour force where future plants could be sited. In India such regions should have at least 45% men with complete basic education and place of birth not further than 200 km. The population should be homogenious from the nationality point of view. Due to the fact that most of the regions concerned are scarcely populated and remote, it is necessary to assume that a research of this kind would be very time-consuming. It ought to be carried out therefore in appropriate advance.

In developing countries of large territorial dimensions, such as India, and a high absolute and relative population growth, a detailed research of labour force preceding final plant location would no doubt bring great advantages to the national economy.

Megee M.
(USA, Joplin, Missouri)

ECONOMIC AND ENVIRONMENTAL MANAGEMENT-REGIONAL
AND ECONOMIC PLANNING AT MID-DECADE

Awareness of the need for planning in order to bring order into the physical surroundings in regions is fairly recent in most parts of the United States.

This paper summarizes the more important findings of a longer study carried out in the United States by the writer under a federal U.S. Department of Housing and Urban Development contract. It involved the detailed inventory and analysis of existing land use and the identification of major regional problems. It includes the first application on a large scale of the USGS Land Use Data and Analysis (LUDA) Program, which will provide a systematic and comprehensive collection and analysis of land use and land cover data on a nationwide basis. In this report U-2 and other high altitude remote sensing materials are used. Computerized forecasts are also generated in the process of developing three alternatives for a regional plan.

Most research and applied work in the United States in the field of planning has been oriented more toward urban rather than toward regional areas. This paper, therefore, represents an initial effort in the field of actual regional planning.

The importance of this study can be summarized as follows. It is the first major regional plan in the United States. Moreover, it involves both urban and non-urban areas, rather than a metropolitan region, as is customarily used in most planning studies. Second, it is significant in that it relates to current U.S. policies in the field of environmental management, and is tied to actual national and regional goals, objectives, standards, and policies. Third, it involves the implementation of policies of the U.S. Department of Housing and Urban Development's 701 program. Last, it is a computerized approach which involves the presentation of three alternatives rather than one for future growth. As such, this document of development is concerned with the uses of land and with a pattern for settlement for the larger future population.

The focus of the paper is on the existing and future state of the natural and man-made physical environment of the region. It emanates in part from the necessity of coordinating the development programs and construction activities undertaken by a number of state and federal agencies, local governments and private enterprises. Decisions made at all levels of public and private enterprise influence the regions' development. Some decisions directly influence settlement, which is further influenced by legislative, administrative, and fiscal means. These decisions then indirectly influence additional decision by local and federal governments, as well as by private enterprise.

A main function of the 1975-2000 Regional Development Plan with accompanying tables, charts, maps, and other materials is to provide a general, but realizable statement of policies and objectives of the region's development. This guide provides (1) comprehensive, long-range standards with which land assembly and development projects must comply for state approval; (2) guidelines for all public facilities development, particularly for the capital improvements program; and (3) guide-lines for the formulation of federally-assisted project plans which must conform to the general regional development inventory and plan.

Those topics selected from the longer report for presentation at the session are concerned with regional environmental management, which research objectives are to improve the quality and availability of socio-economic and ecological information for environmental decision-making; development of improved methods for predicting land use and other secondary consequences of technological and economic developments and alternative environmental policies; and the development, synthesis, and testing of strategies for management of regional resources; and the design and evaluation of the effectiveness of selected technologies for dealing with regional environmental problems.

Pecora A. - Enrico Gregoli F.

(ITALY, Torino)

SUR QUELQUES ASPECTS DES INSUFFISANCES STRUCTURELLES DE
L'AGRICULTURE ITALIENNE: PETITES EXPLOITATIONS ET EXPLOI-
TANTS A TEMPS PARTIEL

1. La concurrence toujours plus forte sur le marché des
produits agricoles et la nécessité qui en découle de rendre
les entreprises rurales plus productives - par la spécialisation des cultures, ou par la réduction qui s'impose des coûts
de gestion, grâce surtout à la mécanisation des travaux et à
la motorisation - ont depuis longtemps provoqué une crise chez
un très grand nombre d'exploitations. Les entreprises frappées
par cette crise présentent trois caractéristiques fondamenta-
les: faible superficie cultivée, cultures non spécialisées,
manque de moyens financiers. Il s'agit donc de petites exploitations, en général inférieures à 5 ha, encore en partie adon-
nées à une agriculture de subsistance, peu ouvertes au marché
des produits agricoles, et dans l'ensemble non comptabilisées.
En dix années seulement, de 1961 à 1971, leur nombre a dimi-
nué de 546.000 unités, c'est-à-dire de 17%. Cette contraction
eût été certainement plus substantielle, si n'avaient pas été
créées environ 140.000 nouvelles petites entreprises dans les
zones touchées par la réforme agraire, si d'autre part beaucoup de cultivateurs n'avaient pas opposé de résistence au
choc psychologique consistant à se libérer du petit morceau
de terre hérité de leur parents, si surtout ne s'était fortement développée au cours des deux dernières décennies, chez
les petits cultivateurs, une tendance à s'adonner, parfois
de préférence au reste, à une activité seconde, en dehors des
limites de leur propre ferme. Cependant, il apparaît que les
petites exploitations seront de moins en moins nombreuses, et
cela dans de fortes proportions: en effet elles représentent
encore les 3/4 des entreprises agricoles nationales, exacte-
ment 2.718.000 sur un total de 3.591.000. Le phénomène connaît
une diffusion véritablement nationale: il est présent partout,
marquant profondément aussi bien les régions où la petite pro-

priété à exploitation directe est la forme la plus importante de lá possession de la terre et de l'organisation de l'espace agricole, que les régions où elle se concentre autour
des villages et des bourgs, et apparaît comme immergée dans un
espace rural, qui peut être autant la grande propriété lati-
fundiaire du Sud que la "ferme-usine" de type capitaliste moderne du Nord. C'est dans ces régions seulement que la petite
exploitation représente moins de 50% du total des fermes et
dans quelques communes à structures agraires très avancées,
telle la plaine du Bas-Milanais, elle a complètément disparu.

De la carte de la distribution des petités fermes (inférieures à 5 ha) et de l'activité extra-agricole des agricul-
teurs - qui a été construite sur une base communale, à
l'échelle de 1/750.000^e, puis réduite à l'échelle de
1/1.500.000^e en plusieurs couleurs - il ressort clairement
que les petites entreprises connaissent une diffusion généra-
lisée, même s'il existe des écarts sensibles sur le plan ré-
gional: úne telle diffusion indique explicitement que les insuffisances structurales de notre agriculture concernent pratiquement tout le pays. Il n'y a pas donc de quoi s'étinner
si dans toute la nation un nombre consistant d'exploitants
agricoles se livrent préférablement à des activités en dehors
de leur ferme: il s'agit là de 1.177.000 fermiers, correspondant au tiers de la totalité des fermiers, parmi lesquels
830.000 ont une activité directement non agricole, ce qui si-
gnifie que cette activité s'exerce dans l'industrie, le commerce et les services.

2. Les exploitants qui retirent la plupart de leurs reve-
nues d'une activité non agricole sont bien nombreux dans certaines zones, fort soulignées sur la carte: d'une façon assez
compacte dans la zone pré-alpine - du Piémont jusqu'aux Véné-
ties - et dans des zones espacées et moins étendues, localisées entre la dorsale des Apennins et la mer Tyrrhénienne -
surtout en Toscane, en Latium et dans une bande qui s'étend
de la Campanie méridionale à la Calabre septentrionale - ces
exploitants devancent le 30% du total. A cause de la différence remarquable des conditions économiques et sociales des

régions intéressées, il s'agit donc d'un phénomène complexe,
qui se manifeste de façon variable dans les différentes parties du pays.

La nouvelle figure de l'agriculteur à temps partiel se
caractérise - par opposition à sa figure traditionnelle, laquelle est désormais presque partout désuète, lorsqu'au travail agricole comme activité fondamentale il unissait un travail presque toujours artisanal durant la "morte saison" des
champs-par au moins trois éléments particuliers: sa double
activité s'applique à des lieux différents; son activité agricole est devenue secondaire, c'est un complément, d'ail-
leurs intégré, de son nouveau travail; ce dernier, enfin,
s'exerce dans les secteurs économiques les plus différents,
dans l'agriculture même, auprès d'exploitations plus modernes,
ou encore dans l'industrie, le commerce ou les services. La
prédominance d'un de ces nouveaux secteurs est évidemment en
rapport avec la structure économique et sociale des diverses
régions: là où prévaut, comme champ d'activité, l'agriculture
même, tel est le cas du Sud, on doit s'attendre logiquement à
une plus grande résistance des petites fermes économiquement
faibles, aidée en cela par la proximité des deux lieux de travail; là où, au contraire, s'impose l'industrie, comme en général dans le Nord, la solution de la crise des petites exploitations agricoles ne peut consister qu'en leur graduelle disparition, car l'agriculteur à temps partiel a tendance à se
fixer où se trouve sa nouvelle occupation: une réserve s'impose en ce qui concerne les générations les plus jeunes, du
moins celles attachées à la petite ferme reçue en héritage et
au travail agricole considéré une détente et un moyen d'échap-
per aux névroses contractées en usine ou en ville. Ceci est
un phénomène courant dans les pays industrialisés de l'Europe
occidentale, et dont il est possible de voir un nombre crois-
sant d'exemples en Italie. Si le travail en dehors de l'exploitation familiale, enfin, touche au commerce et aux services, pour le compte d'autrui on à son compte, une fois encore
le lien avec la terre peut être conservé. Il s'agit en effet
de travaux saisonniers, qui, presque toujours, coincident avec

288

la morte saison des travaux agricoles, et auxquels, par conséquent, peut utilement se consacrer l'agroculteur à temps partiel. La même complémentarité entre les deux occupations se
retrouve chez les émigrants saisonniers à l'étranger, et dont
la plupart résident dans le Midi; ils contribuent dans une mesure sensible - bien qu'on ne dispose pas de données à ce sujet - à rehausser la valeur du travail extra-agricole dans
nos régions économiquement peu développées, et par conséquent
incapables d'offrir autre emploi qu'une activité agricole.

3. Sur la carte qui sera présentée au Congrès, on a aussi
fait figurer l'incidence du travail non agricole effectué par
les exploitants rurales en dehors de leurs fermes: on a mis en
évidence d'une manière significative les régions où se manifeste le plus intensément le phénomène (+30% des esploitants).
On a d'autre part volontairement laissé de côté le travail en
dehors de la ferme, mais de type agricole, pour souligner -
- même de façon indirecte - les zones dans lesquelles l'agriculture manifeste les plus grandes insuffisances dans le domaine des occupations.

L'examen comparatif des deux phénomènes révèle une remarquable variété de situations, qu'on peut en substance réduire
à trois:

a) Coincidence de hautes valeurs pour les deux phénomènes. Un
grand développement contemporain de l'agriculture à temps partiel et de la petite exploitation interesse une ample zone
septentrionale, qui s'étend du Piémont au Frioul, où la concentration industrielle est intense; ensuite, elle concerne
une succession de bassins entremontains et de couloirs de vallées, situés entre la dorsale des Apennins et la mer Tyrrhénienne, où le travail est offert parfois par l'industrie, ou
plus souvent par une maigre activité commerciale et artisanale, qui offre plus l'aspect de la sous-activité que celui
d'un véritable travail.

b) Coexistence de hautes valeurs des petites entreprises et
de basses valeurs du travail extra-agricole. On peut individualiser ici au moins deux situations différentes et même opposées: d'une part la zone d'agriculture privilégiée et de

haut rendement, et de l'autre la zone où manquent les débouchés nécessaires dans d'autre secteur économique, comme dans le Sud.

c) Coïncidence de basses valeurs pour les deux phénomènes. Elle se manifeste dans un grand secteur de la plaine du Pô, où sont plus nombreuses les moyennes et grandes exploitations agricoles, douées de structures productives solides.

Il faut enfin souligner que les données du travail extra-agricole par nous utilisées n'offrent pas un cadre complet du phénomène: elles indiquent concrétement seulement le nombre des exploitations en crise. Si on considère aussi les aides familiales, un calcul approximatif peut facilement faire monter le nombre de ceux qui vivent en partie de leur ferme et en partie surtout d'un travail extra-agricole, au chiffre vraiment remarquable de à peu près 1,3 million; lequel monte à presque 1,8 million, si on tient compte aussi des entreprises dont les membres travaillent auprès d'autres exploitations rurales.

IX. HISTORICAL GEOGRAPHY

Brunger A.G.
(Canada, Peterborough)

THE IMPORTANCE OF KINSHIP IN THE SETTLEMENT PATTERN OF AN AREA OF EARLY ONTARIO

The paper demonstrates the importance of the social factor of kinship in the development of the spatial pattern of settlement by individuals during the initial European occupation of Ontario in the early nineteenth century. The area considered is in the south-western peninsula of Southern Ontario, formerly Upper Canada, in the centre of the northern shore-line of Lake Erie. The period of settlement extends from 1803 to 1818 - the first fifteen years of the colonisation scheme known as the Talbot Settlement.

The spatial pattern of settlement in the Talbot Settlement was reconstructed from primary records and attempts were made to explain it in terms of traditionally important factors such as access distance to major route-ways, foci of economic importance and the quality of the land for agriculture. These were not as significant as in other areas of early Ontario and other factors were introduced including the official control of Colonel Talbot, the head of the Settlement, and that of kinship. Kinship was measured, in fact, by means of nominal census records only and only agnatic links within a family were used, as a result. Individual immigrant settlers with kinship links of this type were expected to select a settlement location as close as possible to their relations - the latter having either preceded them, or accompanied them, to the area.

Detailed records of individual settlement enable researchers to identify the 200-acre lot of land selected by
the settler and, at worst, the year of arrival. In the majority of cases the month and year of settlement is known. An
impression of the spatio-temporal sequence of settlement may
be created which permits the gauging of the importance of the
kinship factor in the process. Data exist on 537 individuals
who had settled by the middle of 1818. 208 of these shared
eighty-three different surnames. A high proportion of settlers occupied land within a distance of the two nearest
available lots at the time of settlement. The paper concludes
that kinship had an important influence in settlement initial-
ly owing to the large proportion of total settlers apparently
locating in response to this factor.

Vozovik Yu.I.
(USSR, Moscow)

SHORT-TERM CLIMATIC FLUCTUATIONS AND ECOLOGICAL FORE-
CASTING

There are two aspects which should be distinguished
within the forecasing problem - the ecological and geographic
ones, the former being the forecast of environmental conditions, the latter - the prediction of landscapes, ecosystems
and other phenomena resulted from the realization of the conditions. Therefore the geographical forecast is a part of the
ecological one.

At the elaboration of ecological forecasting one takes
into account tendency (or tendencies) of the climate development and gives grounds for the ecological consequences of
the climatic tendency. Geographical forecasting needs determination of expected duration of climatic epochs which essentially complicates the problem. Before the scientific-tech-
nological revolution the climate development was controlled
only by natural factors. Quick advance of industry caused
more and more considerable breaks in regime of natural pro-

cesses not only at regional but at global scale as well. The climate is influenced most clearly (through heat influx, contamination and changes of gas composition in atmosphere etc.). Various participation of the factors and natural fluctuations of solar energy may result both in increase and in decrease in energetic budget of the Earth; both variants must be taken into consideration.

Assuming the ecological conditions to be determined with amount and distribution of heat and moisture and the conditions realization to depend on their duration, the task of forecasting (from the discussed point of view) is to estimate the energetic budget evolution, as the value of the budget controls the fields of temperature and humidity structure.

The value and regime of humidity are known to be conditioned by western transfer in troposphere and to be result (direct or indirect) of high frontal zones (HFZ) dynamics. There are several HFZ, the most important ones being those between arctic and polar air masses and between polar and tropic air masses. These HFZ change their position through a year from high latitudes in summer to low latitudes in winter. Thus the HFZ position depends on temperature field and the latter results from amount of the solar energy. There exists a certain structure of temperature field corresponding to each energetic level, and structure of humidity field corresponding to each temperature field. As the correlation can be identified at the annual cycle, we can assume its existence for periods of longer duration, i.e. for climatic changes. Some studies (Sergin, 1974) confirm the possibility of such extrapolation.

On the base of the correlation a theoretical model is developed. Varying the model under conditions of nonstationary energy influx (with given coefficient of variation) a series of climatic epochs may be obtained within the limits of a given extremum. Using paleoenergetic data with adequate degree of detalization it is possible in general to date the epochs. Such is the task of retrospective approach to the ecological forecasting problem which can be solved by theore-

tical modelling of the climate dynamics and tested by paleogeographic reconstructions. When an acceptable correlation is
achieved between the theoretical model and paleogeographic
reconstructions there is a real possibility of long term climatic (and therefore ecological) forecasting. Time scales of
the retrospective analysis and forecasting have to be commen-
surable, the epochs in question being about one-two hundred
years long. The time scale is used at paleogeographic studies
within the historic period, from 5-6 thousand years B.P. up
to now (Markov, 1955). The small time interval needs a special technique of selection and interpretation of primary information, proper time and space coordinates being a necessary condition.

Consideration of several epochs of the historic time
reveals the amplitude of the climatic fluctuations to be of
the same order as those of Holocene and Pleistocene stages.
Because of the short period of the former they were not realized in landscape; only the most labile elements had time
to respond the fluctuations, more conservative ones didn't
change, being ecological relics of previous epochs.

Using the same time scale in the estimation of future
events development we meet the same feature. At large scale
only ecological conditions controlled by climate are fore-
casted, the degree of their realization in landscape depends on
the climatic epochs duration.

Some preliminary results of the method application to
the USSR territory are as follows.

It is to be noted that no general climatic tendency has
been revealed at the whole territory, the direction of changes
being different at each thermic belt and province of humidity.

Thus a further increase in the Earth's energy budget
will considerably influence the structure of humidity field.
Everywhere at the North of Eurasia precipitation amount will
be increased. Some ecological conditions will be created for
forest movement towards higher latitudes (dark coniferous
forest - at the western part of subarctic belt, mixed forests
at Western Siberia, larch and pine forests at Yakutia, light

294

coniferous forests – at the North-East of the USSR). The intensity of termafrost degradation will be increased, especially at the North of Europe.

There will be general increase in temperature indices within the northern half of the temperate belt, partly due to the increase in heat budget, partly due to increase in radiation index of dryness because of decrease in normal humidity, especially at the Atlantic sector. The ecological consequences will be: steppization of central regions of Russia, replacement of dark coniferous to mixed forests at the northern regions. The tendency will be even more pronounced at Western and Eastern Siberia. At most part of the Far East taiga will be replaced with mixed forests and at the South steppe-forest ecological conditions will come into being.

An optimization of ecological conditions will take place at the southern part of temperate belt due to increased precipitation and temperature index. An ecological areal of broad-leaved forest will be enlarged, forest-steppe associations extending its areal eastward. Area of dry steppes and semi-deserts will decrease, ecological conditions of steppes will be created. Deserts will be replaced with semideserts. At the South of Primorye there will be conditions for gradual invasion of subtropic forms. It will be drier at subtropic belt but it will be ecologically pronounced only at the East of Transcaucasian region and at Middle Asia foothills.

Changes in general tendency of energetic budget of the Earth seem less probable. In such case the climatic situation will be similar to that of XVIII – first half of XIX centuries. The climate of the time has been studied well enough so the paper does not discuss the consequences of such development.

It is necessary to say in conclusion that the results are but preliminary. A great deal of work is still to be done for the model development as well as for its experimental test (by means of paleogeographic reconstructions).

Imbrighi G.
(Italie, Rome)

POUR L'ETABLISSEMENT D'UN PROGRAMME DIDACTIQUE
DE LA GEOGRAPHIE

Tout ce qui a fait l'objet de travail individuel ou
d'équipe, de recherche documentée en vue de la rédaction de
mémoires de géographie - même quand ce travail n'a pas été
approfondi - a contribué à développer de multiples intérêts
culturels qui, pour résumer, peuvent être classés ainsi:

1) Préliminaires géographico-physiques sur la région;
2) Etablissement humain dans les Abruzzes;
3) Perspectives économiques du territoire;
4) Thèmes et principes de Géographie appliquée.

Les préliminaires géographico-physiques sur la région
des Abruzzes ont permis, entre autre, de parfaire les connaissances en seismologie de la région, dans le but d'élaborer
une cartographie spécifique et également de rédiger un essai
sur la spéléologie régionale, un commentaire critique sur les
indices pluviométriques de la région, de grouper les observations géographiques sur l'aspect hydrographique du territoire
et sur les phénomènes des calanques et des éboulements.

Des études particulières ont permis l'installation de laboratoires de météorologie hypogée dans le Massif de la Maiel-
la; ces laboratoires, qui utilisent des instruments technico-
-scientifiques modernes, se sont par la suite perfectionnés.
Nous pouvons mentionner également d'autres études, riches
d'observations intéressantes, qui ont été faites sur les con-

séquences phytogéographiques et sur le climat du territoire-
-témoin choisi pour les expériences.

Bien que la plupart des étudiants de la Faculté soient
relativement peu préparés en physique naturaliste, puisqu'ils
sont principalement concernés par des études historico-huma-
nistes, la documentation qu'ils ont réunie a fourni des données intéressantes et des éléments de première importance qui
pourront être utilisés pour des approfondissements ultérieurs.

Les étudiants ont nettement contribué à éclairer certains
aspects de la géographie concernant l'établissement humain
dans la région-témoin. Leur rôle a vraiment été utile pour
des raisons énoncées par beaucoup et qui consistent à quali-
fier d'humaine la géographie moderne.

Très intéressantes également les études faites sur les
conséquences socio-économiques de l'assèchement du Fucino;
sur les rapports Parc National des Abruzzes - population;
sur les aspects de l'émigration dans d'autres régions d'Italie et dans les pays européens et extra-européens; émigration
étendue à toute la région ou limitée à un centre habité bien
défini; sur les variations démographiques de la période
1861-1961 dans les différentes zones du territoire; sur les
centres les plus élevés de la province et sur ceux qui ont
disparu sur l'ensemble de la région; sur l'évolution de l'urbanisme et de la démographie des centres habités les plus im-
portants du territoire-témoin; sur les critères relatifs à la
détermination des centres satellites.

Les perspectives économiques du territoire, en rapport
avec les ressources agricoles, minérales et hydriques, ont
particulièrement attiré l'attention des chercheurs sur ces
préliminaires, que ce soit dans le but de faire naître d'autres activités industrielles ou pour renforcer ou spécialiser
certaines branches de la production artisanale, pour donner
une impulsion nouvelle au développement économique et social
de cette zone.

Un groupe a dirigé ses recherches, dans le domaine de la
géographie des transports, sur la variation des écosystèmes
du territoire à la suite de la création d'infrastructures
aéroportuaires, routières et ferroviaires.

Toutes ces recherches, ainsi que toutes celles dont nous n'avons pas pu parler - ayant été présentées et examinées et donc disponibles - contiennent des chiffres et des données mis à jour au moment de la préparation des mémoires, ainsi que des commentaires critiques, c'est pourquoi elles ont beaucoup intéressé la Députation locale d'Histoire de la Patrie qui a demandé officiellement à l'Université l'index bibliographique de ces travaux pour pouvoir l'ajouter aux quelques 400 fiches qu'elle possède déjà.

Les problèmes de la Géographie médicale sont nouveaux et intéressants: celui des jumeaux, par exemple, dans les quatre provinces; ou encore certaines observations sur la "Leishmaniosi cutanea" qui se répand actuellement sur la région du littoral et sur certains cas de mortalité; ces observations sur la mortalité ont été faites avec l'intention, de la part des savants ou des organismes compétents tel l'Organisation Mondiale de la Santé, de découvrir s'il existait un rapport entre les manifestations d'un mal aujourd'hui incurable et certaines conditions ambiantes, psychiques, climatiques, alimentaires, ethniques ou sociales.

Nous pouvons également citer, toujours dans le vaste domaine de la Géographie appliquée, les questions géo-politiques qui ont été étudiées par le biais de la philatélie d'après les résultats électoraux; ou encore des problèmes examinés par la CEE ou les singularités frontalières des provinces et des communes.

En conclusion, en regard des nombreuses recherches reflétant des intérêts culturels particuliers ou individuels ou de simples indications qui ont été approfondies à titre d'exemple, est élaboré un programme d'une série d'essais sur certains points essentiels des sciences humaines, comme justement le Tourisme et l'utilisation des loisirs, autour desquels doit se concentrer l'attention des chercheurs, selon l'esprit des conceptions modernes qui en font une science d'observation subtile, de confrontation documentée, de synthèse critique qui s'appuie sur des bases historiques et des conditions nécessaires naturelles; science qui reste sensible, comme peu le sont, aux facteurs sociaux et politiques du monde moderne.

Machyček J.
(ČSSR, Olomouc)

CURRENT PROBLEMS AND TASKS OF SCHOOL GEOGRAPHY IN THE CZECHOSLOVAK SOCIALIST REPUBLIC

The problems and tasks of school geography in ČSSR have at present been discussed at a number of universities and university colleges educating teachers of geography, at geographical institutes of the Czechoslovak Academy of Sciences, and pedagogical research institutes both in Prague and Bratislava. The Czechoslovak Geographical Society of the Czechoslovak Academy of Sciences in Prague has in the last years emerged as the coordinating and directing organization. The 12th Congress of the Society held in July 1972 in Ceske Budejovice passed a resolution instructing the Central Committee of the Society and its Commission for School Geography to concentrate its activity till the next Congress planned for July 1975 in Plzen particularly on the following tasks:

1. To make sure that geography as a subject has its adequate position in the curricula and teaching programmes of Czechoslovak schools so that an effective system of teaching geography and developing the pupils' skill is guaranteed, and to elaborate the respective, well-founded proposal to be sent to the Ministry of Education.

2. To direct the Society's attention to modernization of teaching geography with the aim of getting as close as possible to the present state of geography as a science, and to establish cooperation with the respective institutions. The elaboration of a schedule popularizing geography is part of this task.

3. To negotiate with universities and the Scientific Collegium for Geology and Geography of the Academy and with the Pedagogical Research Institute to have a planned set of modern textbooks and teaching aids worked out for all the grades and types of schools.

The mentioned items of the resolution of the 12th Congress of the Society have become a concrete guide for the work of the Commission for School Geography and for the Expert Group for School Geography of the Society's Central Committee in between the two congresses, in a period when the definitive solution of problems associated with the rebuilding of individual cycles of Czechoslovak education is being looked for. For the first time in the Society's history there is a chance for it to influence directly the elaboration of the curricula, teaching programmes, textbooks, and teaching aids for instruction in geography, and for the unification of the still scattered effort of individual geographic institutions to attain the common goal - all-round modernization of the teaching of geography on all grades of our schools. In view of the large extent of work required in the solution of this task, four separate sections were established at the Expert Group for School Geography to tackle the problems how to modernize the content of geography as a teaching subject, the political and educational aspects of its teaching matter, modernization of teaching technique and of methods and textbooks.

The conclusions derived from the work of the Expert Group for School Geography in the period between the congresses are such that their realization would help much in removing the drawbacks found in the present state of teaching geography, considering the planned rebuilding of the general educational and vocational schools necessarily involving university colleges training teachers. The conclusions are:

1. To modernize the teaching matter of secondary school geography in accordance with the present state of geography as a science. The expert group has elaborated criteria this modernization should follow. For illustration it was possible to present the curricula of geography used at Soviet secondary schools including both systematic approach to landscape study and theory of geosystems.

2. To pay attention to the remarkable both teaching and educational importance of geography as a subject considered

300

among the educational tasks of generally educating and vocational schools, the substantial share it has in the acquisition of the scientific view of the world by the pupils, in the education to socialist patriotism and internationalism. To stress the integrating role of geography in the curricula of secondary schools. This structure is indispensable for understanding economical and political relationship governing the world and for the acquisition of the scientific view of the world by the pupils.

3. To make sure that the teaching of geography at general educational secondary schools is given such a position and number of lessons a week that would enable the teachers of geography to make use of all the educational and instructive posibilities this subject offers. At the same time, to try to keep continuity in teaching geography through the whole duration of the secondary school attendance.

4. To introduce the teaching of geography into all vocational schools in the extent and modification corresponding to the pursued specialization of individual types of these schools, and at least to some selected establishments training apprentices.

5. To reorganize the present single course postgraduate study of geography into a system of regular, short-time postgraduate courses involving both methodical and professional instruction at the teacher training colleges of the universities and at the independent pedagogical faculties.

6. To intensify the teaching of the didactics of geography at university colleges educating the teachers of geography. To strengthen particularly the practical component of the teaching programme so as the provide the graduates from these type of schools going to teach with the required knowledge of methodology and skills necessary for further realization of the modern conception of teaching geography.

7. To publish modern textbooks of geography satisfying all the demands of teaching geography at the present standard, and the respective methodological manuals for teachers accompanying the textbooks for individual classes. To reach

quicker circulation of these textbooks so as not to allow their coming out of date.

8. To devote full care to the production of teaching aids for instruction in geography and to make sure that the best prototypes are manufactured so as to serve the modernization of the teaching techniques as soon as possible.

9. To pay the same attention as to the didactics of geography in the university education of geography teachers to systematic content modernization of the professional component of the student's instruction.

10. To unify and intensify research in the didactics of geography to be able to realize the preceding items. There is a plan of coordinated research in this scientific discipline elaborated in the form of the research project of the Ministry of Education of the Czech and Slovak Socialist Republic. The rising demands on research in this respect necessarily require more scientific workers in the didactics of geography having the opportunity of obtaining academic scientific, or combined teaching and scientific degrees. This would understandably add to the prestige of didactics of geography as a scientific discipline.

Raik A.A.
(USSR, Tartu)

ON THE EXPERIENCE OF TRAINING SPECIALISTS IN APPLIED GEOGRAPHY AT TARTU STATE UNIVERSITY

Large-scale geographic research of a general theoretical nature is limited in the Estonian SSR and so the majority of the graduates in economic and physical geography will start working at practice establishments among which planning and designing institutions prevail. The preparation for applied activities has some specific features.

The first specific feature in the preparation of a practical worker in the field of geography arises from the essence of applied science. It should not be imagined that its role is confined to the practical interpretation of the achievements of the purely theoretical science and also the preparation for carrying them into practice. Its essence is specific applied research the characteristic features of which in our opinion are the following. If a purely theoretical investigation proceeds from the properties of an object as such and is aimed at studying its genesis, morphology and structure, a practical investigation proceeds from the connections between the object and the sphere where it is used, from a sideblock of the "object-user" system. Applied research need not, however stand inferior to basic research as less complex or less creative. In view of this only such an investigator is likely to succeed whose way of thinking and psychological frame of mind permit him to see problems from the position of specialists in other fields, to see his own object of study with the eyes of an outsider. Such ability to view things with an outsider's eyes should be specially developed in the course of preparing specialists in applied geography.

Another specific feature in the training of a practical worker in the field of geography can be derived from the situation in which he will have to work in the future. Namely, there is no special board to conduct and co-ordinate the work of geographers in the republic as there is the Board of Hydrometeorological Service for graduating climatologists, hydrologists and oceanographers. A geography graduate will often have to start working in a team where he will be the first and only geographer. He must therefore be able to see the geographical aspects and connections of the problem appointed to the team, to independently formulate his problem of investigation and outline the ways, methods and procedures of solving it. He must himself carry out the investigations and pass on his results to the team, so that they may make up a part of the collective investigation. Hence the task to make the future specialist in the course of studies most possibly independent, capable of standing up for his speciality. If this hasn't been achieved, a geographer will degenerate, he will become a mere assistant to his team-mates, specialists in other fields. We may frequently enough observe in the above-described case the conversion of a geographer into a mediocre specialist of some neighbouring field.

From the above it follows that besides theoretical preparation in the fields of general and applied geography and the inculcation of enthusiasm and love for their chosen speciality, it is necessary to impart to geographers as much practical experience as possible in the independent solution of concrete tasks to instil in them self-confidence in their own abilities.

In giving the necessary preparation to future geographers at Tartu State University irrespective of their concrete narrow profile, attention is paid to giving them a solid foundation in mathematics and the use of quantitative methods of investigation, so that they would be able to make use of the computer in writing their diploma theses. Scientific research at the department and all subjects of study generally rest on a systemic basis. In nature studies the principles of ecology and environmental protection are consistently followed.

To guarantee a special preparation in applied geography the students are expected to take a cycle of special subjects from their 3rd to 5th study years as well as the professional practice and to write course-papers and a diploma theses.

Special subjects for students of physical geography include land amelioration and the fundamentals of agricultural land tenure, the theory and methods of the estimation of natural conditions and resources, landscape maintenance, landscape planning, regional planning and geographical informatics. A part of the above-listed subjects have been included in the study plans approved by the Ministry of Higher Education. The plans came into force in 1974 and according to them the subjects will first be introduced in 1976 - 1978.

Applied geography is a very dynamic field where urgent changes are generated by new constant demands in social sphere, by the penetration of geography into new fields of activity and also by the application of new research methods.

From the above it follows that study programmes should be looked through every year and new ideas be introduced into them. By doing this we keep the study plan on the up-to-date level without a frequent need for changing the names of the subjects in the plan. Of ideas thus included in the programme we might mention those of geographical examination, medical geography, amelioration geography.

Professional practice has a special place in the curriculum of the department, as it will place the student in situations which will be as nearly similar to his future work as possible. Sometimes the future specialist may get acquainted with his prospective place of work during the professional practice. The total 27 weeks' length of the practice creates facilities for varied work. A close contact with the establishments the practice is carried out and will ensure that the student will be included in a team engaged in solving a problem which envolves a geographer's task, and which would give the student an opportunity to get acquainted with and take part in all the specific forms of work essential for a geographer. The situation is ideal if the team includes an experienced geographer whose presence will make it much easier to solve the problems that crop up, and will help the student acquire the necessary experience and skills. In an adverse case the more responsible tasks will fall on the shoulders of the supervisor from the university.

The chair will see to it that the student's work will not be narrowly limited to the collection of data and the carrying out of assigned research procedures. The student should be in the know of overall research, its strategy and general methodical conception, of the whole system of the problems to be solved. This will equip him with the necessary qualities to develop a broader outlook and skills required of a leader.

In the acquiring of the experience and skills of independent research work course-papers and diploma theses play an important role. They are closely interwoven with the professional practice being sometimes the preparation for the

305

latter and sometimes its logical sequence interpreting the data collected during it. We are of the opinion that a course paper should not be tied up to any particular subject of study, or else it would unnecessarily restrict its subject-matter.

In the choice of the subject for a course-paper starting from the third year already, we try to keep in mind the future diploma paper. Alongside with general suitability and topicality we try to take into consideration the spheres of interest of a student and see to it that he will have an opportunity of collecting a sufficient amount of initial data. Course-papers and later on diploma theses are naturally forms of processing and generalizing the data collected during the professional practice. The diploma theses of some of the students are closely and directly associated with the research problems of their prospective place of work.

Stroev K.F.
(USSR, Moscow)

TEACHING PRACTICE IN THE SYSTEM OF TRAINING GEOGRAPHY TEACHERS

Teaching practice stands high in the priorities of national planning for the training of geography teachers in the USSR. It plays a bigger part in relating academic work in geography to the professional training of teachers and in linking the education of students with the demands of life than all other ingredients of teacher-training curriculum. It is the touchstone against which students can test their choice of a future profession and it enables them to reveal their attitudes to a teacher's work or to develop their capacities to carry on such work.

Far from having been constant, the claims of teaching practice in the institute's curriculum have followed every development in the rearragement of the general teacher-training system. To discover a structure of teaching practice which would be best suited to the basic needs of teacher-training has always been one of the chief concerns of Soviet pedagogical science as well as of training institutes, whose long record of research and experiment has added considerably to the competence of their staffs. However, there are still many issues relating to matters of organization and structure of practice whose optimal development concerns both educators and geographers. These include, for example, the timing and duration of teaching practice; criteria for choosing schools in which it is to be done; the sequence of stages in carrying out the teaching practice program; relating the work to be done during the teaching practice to the training institute's curriculum on the one hand and the school's syllabus in geography on the other; and, last but not least, come matters supervision.

No one has ever challenged the obviously indisputable view relating the success at teaching practice to the level of student's ashievement in the academic training based on the study of the curriculum specialist geographical subjects as well as psychology theory of education and methods of teaching geography.

To indtroduce students to schools and to develop for them some basic skills and habits in rearing children, a large part of the course in the junior years (which includes the study of such subjects as Introduction to Education, Developmental Physiology and the Schoolchild's Hygiene, Theory of Education, General Psychology) is taken up with practical work at schools. The students learn the basic principles of organizing and carrying on the education of children in schools, the essential of a teacher's work as well as that of a form master. They also observe children in classrooms and in extra curricular activities and do some educational work under the guidance of school teachers.

This practical work in schools (which is also as part of students' public activities) is carried out without discontinuing institute classes.

In the two senior years students do their teaching practice on a full-time basis: 6 and 8 weeks in the penaltimate
and the last year respectively.

Experience shows that the best time for the teaching
practice is the penaltimate year (the third year of a four-
year course or the forth year of the five-year course) is the
third term of the school (four-term) year (after the winter
holidays). For students in their last year the most suitable
time is the beginning of the school year (the first term).
By the time their first teaching practice begins, the students
will have covered all the topics of "The Essentials of Teaching Methods" and "Methods for Teaching Physical Geography"
and will thus be equipped to teach physical geography at
school. Their second practice is done largely in the senior
forms which do economic geography.

In both cases teaching practice is but an extension of
the institute course and students are required to conduct record (test) lessons and carry out extra-curricular activities
both in the capacity of specialist teacher and that of form
master. To do this work each student is attached to a class.

All the work done during teaching practice is closely
supervised by a member of the institute staff specializing in
education and teaching methods.

Once it was proposed that students doing their
last-year practice should be given probationer status underlying assumption that school teachers would be able to give
effective supervision. Soon, however, serious disadvantages
of this proposal were discovered. On the one hand, students
had not yet acquired sufficient knowledge, and on the other,
to fully supervise students' work while attending to his professional duties proved to be too burdensome for a school-
teacher. At present, his role is confined to giving general
advice on his subject.

But with teaching practice closely supervised by a methods chair staffmember it still remains to be decided to
what extent a student should be left to work on his own.
Should all the test lessons be supervised or only some of

them? What kind of material and what amount of it should a student be required to present to his superviser (lesson plans, elaborate outlines of lesson sequences notes of exercises to be done in class, etc.)? There is considerable difference of opinion as to how to solve these questions in a best way under different circumstances. As a rule, the results of test lessons are improved if having been introduced to essential teaching methods, students thoroughly analyse various possible types of lessons during their institute classes in methods and later follow these guiding lines in the course of their teaching practice. In this case it is inadvisable to allow students to conduct some of the test lessons (preferably, the last ones) "tête a tête" with the class.

Schools are attached to teacher-training institutes and serve as bases for conducting teaching practice. If located conveniently near the schools they may also be used for carrying on regular experimental work related to be professional training of students. Such schools must meet the following requirements: they must be staffed by highly competent geography teachers; the educational process must be very well organized; they must have a geography study room, the number of classes must be sufficient for teaching practice purposes.

While preparing for their test lessons, students learn to solve the following problems:

1. What kind and what amount of knowledge is to be imparted to the class?

2. What will be the educational effect of the lesson?

3. What skills and habits will the pupils acquire?

4. What type of lesson is to be prefered to achieve the highest educational effect?

Even as they attend institute classes and methods for teaching geography, students reveal their special interests and leanings in matter of education, - effect which makes it possible to offer them research themes to be elaborated during their teaching practice. These are introduced to serious research in methods for teaching geography. Such themes are made known to students at an introductory conference preced-

309

ing their teaching practice, and they are later expected to
report on the results of their research at final sessions.

To acquire skills and habits requisite for taking children on excursions and hikes and for conducting open-air
classes, students are to do some kinds of educational work
during their field studies (in geography and geology).

Wahla Arnošt
(ČSSR, Ostrava)

AN ANALYSIS OF INSTRUCTIONAL ASSIGNMENTS ON GEOGRAPHY BY MEANS OF AUTOMATIC COMPUTER GUIDANCE

The didactics of geography does not stand aside in making use of automatic computers which have been installed at
Czechoslovak colleges and universities. As far as the didactics of geography is concerned, this technique is expected to
be applied in three main spheres coming into existence:

1. scientific and research sphere

1.1. sphere - geography subject matter - an analysis of didactic material, an automatic generating of assignments for geography instruction, elementary
subject matter, intersubject relations ....

1.2. sphere - a geography teacher - an analysis of the teacher's work in the instructional process.

1.3. sphere - a student in geography instruction - an analysis of the student's work in the geography lesson, an analysis of the student's performance.

1.4. sphere - a theoretic one - modelling instructional process, investigating the system "teacher - instructional automatic machine", "student - instructional automatic machine", investigating
the process of geography instruction in these
systems (levels): an automatic computer, one
student, a group of students, a class, a school,
a group of schools.

2. information sphere (pedagogic data banks)
2.1. the sphere of scientific and research data - good for
checking, storing and processing data on research of geography subject matter, a geography teacher's personality,
and of a student in geography instruction, on research
of the way how geography fulfils tasks in the process of
formation of a socialist man.
2.2. sphere - establishing connection on to the school information network
2.3. sphere - establishing connection on to the geographic
information network
3. instructional sphere
3.1. an automatic computer as a teaching-aid at colleges
training teachers on geography.
3.2. an automatic computer as an assisting tod for geography
instruction - functioning as a certain source of information (a mechanized reference library, geographic encyclo-
cyclopaedia, a means for practising, establishing and
testing the subject-matter.
3.3. an automatic computer as a means of management during
lessons - the instructional process is held as a direct
communication of the student with the computer via the
terminal.

Let us have one practical instance:

Automatic computer and analysis of didactic material

The author of the present paper investigated instructional
assignements in textbooks on geography. In a vast majority of
contemporary textbooks published both in Czechoslovakia and
abroad, there are instructional assignements, that is questions, exercises, tasks. Data processing on instructional assignments were carried out from a collection of geography
textbooks: 5 basic-school level manuals, 3 high-school level
ones, 1 university level manual.

The data were prepared with the aim at executing the following types of analysis of instructional assignments on
geography:
a) analysis of frequency - that is, determining a number of

all assignements in a chapter, a textbook, a corpus; a number of operations to be performed by the student so as to solve the exercise; an average number of assignements within one lesson; a number of revision assignements, a number of assignements for practising.

b) analysis of formulation - that is, determining the linguistic aspect of instructional assignements
c) analysis of operation - that is, determining didactic purposes of respective assignments, stating the subject of student's operations at working out the assignments, what operations (intellectual operations) does the instructional assignement raise on the students' part?

The analysis of frequency gave evidence that both the number of assignements in a textbook and the average number of assignements in one lesson were due to the approach of the author. As for this sphere, the authors of textbooks are not limited by any instructions. In substance, as far as the number of assignements is concerned, there seem to be no trends whatsoever. The research confirmed the necessity of paying more attention to frequency of instructional assignements in geography textbooks and of attaining a certain optimization.

The analysis of formulation uncovered two categories of instructional assignements - the assignements formulated as inquiry and those formulated as instruction. Morphologic analysis resulted in a collection of action verbs used by authors in formulating assignements as instructions. Syntactic analysis dealt with the whole formulation of the assignement.

The analysis of operation focussed the following categories of instructional assignements: those requiring operations with notions, numbers, maps, picture and graphic material and with apparatus. Absolutely highest frequency was achieved in assignements requiring operations with notions, then came those requiring operations with maps. The analysis of operation was based on taxonomy of instructional assignements as conceived by D. TOLLINGER. This taxonomy consists of five categories of instructional assignements: 1. assignements requiring reproduction of knowledge, 2. simple intellectual operations, 3. complex intellectual operations, 4. conveying information, and 5. creative thinking. Each of these categories is further subdivided into three to nine degrees.

At preparatory stage for data processing by means of automatic computer TESLA 200 each instructional operation was first labelled with a number and then encoded.

The taxonomy as applied here proved to be a good method for the exact way of gathering data on instructional assignements in textbooks on geography.

Knowledge of operational value of instructional assignements is of great importance for the educational work of teachers at all school levels.

The present method is not only a contribution to a special theory of instructional assignements (namely on geography) but also a contribution to the general theory of instructional assignements and to the theory of textbooks.

The example shows that the progress in didactics of geography in CSSR can be helped also by means of computers. The didactics of geography as developing mainly at colleges and universities has already all technical means at its disposal.

A

Abrahams A.D. (I-121)

Abramov L.S.(X-110)

Abramova T.A. (I-350)

Acevedo Gonzales M. (I-125;
                              X-107)

Adams R.B. (XII-171)

Adams W.   (II-171)

Adejuyigbe O. (VIII-69)

Agafonov N.T. (VII-149)

Aganbegyan A.G. (XI-53)

Aguilera H.N. (IV-158)

Ahnept F. (I-248)

Aitov N.A. (VII-217)

Aivarian A.D. (XII-98)

Aizatullin T.A. (III-10)

Akramov Z.M. (VI-76;
                      VIII-73)

Aksakolova G.P.(X-II)

Alam S.M. (XI-98)

Alampiyev P.M. (VI-19)

Alanen A.R. (VIII-75)

Alayev E.B. (VI-16)

Alekseyev A.I. (VII-350)

Alexander C.S. (XII-188)

Alexeyev B.A. (V-103)

Alexandrova T.D. (V-67)

Alexandrovsky A.L. (I-402)

Amiran D.H.K. (VIII-78)

Ananieva L.M. (II-17)

Antonov B.A. (I-20)

Anisimova J.N. (VI-177)

Aristarkhova L.B. (I-252)

Arkhipov Yu.R. (VI-270)

Armand A.D. (XI-130)

Armand D.L. (XI-130; V-131)

Arnold B. (X-58)

Aseyev A.A. (I-23; I-128)

Astahov V.I. (I-255)

Ata-Mirzaev O. (VII-307)

Aubin M.L. (X-15)

Avdotyin L.N. (XI-103)

Avello O. (XII-71)

Ayoade J.O. (II-20)

Azbukina E.N. (I-89)

Ayyar N.P. (XII-192)

B

Babichenko V.N. (II-65)

Bǎcanaru I. (VII-224)

Bacauanu V. (I-131)

Bach O.W. (II-141)

Bachenina N.V. (I-252)

Bachvarov M. (VI-37)

Badea L. (I-258)

Baeva R.V. (I-89)

Baklanov P.A. (V-62)

Balauta L. (IV-72)

Balteanu D. (I-133)

Andan O. (VII-219)
Anderson J.R. (VIII-82)
Andreev N.V. (X-II)
Andrianov B.V. (IX-9)
Anisimov V.I. (I-262)
Annenkov V. (IX-13)

Barkova E.A. (VII-153)
Barr B.M. (VIII-84)
Barsch H. (V-135)
Bartaletti F. (VII-248)
Barth L. (X-17)
Barths I. (X-61)
Bartkowski T. (VII-225;
           V-78)
Barykina V.V. (XII -107)
Bassols-Batalla A. (VIII-86)
Bastie J. (VII-230; VI-186)
Battistella R.(VIII-90)
Bauer L. (II-312)
Baul A.F. (VI-84)
Baumann D. (VII-232)
Baumgart-Kotarba M. (I-27)
Bazilevich N.I. (IV-62;
              IV-192)
Beaujeu-Garnier J. (VII-219)
Becsei J. (VI-191)
Beguin H. (VII-236)
Beklemishev K.V. (V-116)
Belanger M. (VII-236)
Belchansky G.I. (VIII-234)
Belec B. (VI-194)
Beluszky P. (VII-100)
Ben-Arieh I. (IX-39)
Benedict E. (VI-274)
Bennion L.S. (VII-61)
Bentley J.R. (XII -41)
Berenyi I. (VIII-238)

Ban N. (VII-152)
Bandman M.K.(VI-80;VIII-15:,
           XI-53)
Bannikova I.A. (XII -101)
Barbieri G. (VIII-230)
Barbina G. (VIII-220)
Berzina M.Ya. (VII-102)
Bezrukova T.P. (V-106)
Bilello M.A. (II-23)
Billard A. (I-319)
Biondi G. (VIII-93)
Biosca Acilera L. (I-276)
Bird I.B. (III-64)
Birina A.G. (I-402)
Bladen W.A. (VII-277)
Blagovolin N.S. (I-23)
Blazek M. (XII -168)
Blazhko N.I. (VI-270; X-64)
Blouin C. (VIII-57)
Bodenkov Yu.P. (V-62)
Bogdan O. (II-318)
Bohra Dan.Mal. (XII -196)
Boislaroussie J. (VII-219)
Bojoi J. (I-384)
Bandyopadhyay M.K. (XII -9)
Borissov Z. (VII-353)
Borlenghi E. (VIII-97)
Borsy Z. (I-134)
Botcharov Yu.P. (VII-237)
Boudyko M.I. (XI-26)
Bourian A.P. (VI-109)
Bourne L.S. (VIII-99)
Bours A. (VI-88)
Boytel F. (XII -41)
Braithwaite R. (II-274)
Bravard I. (I-30)
Brawer M. (XII - 201)

Demek J. (V-11)
Demko G. (VII-22)

Dynowska I. (II-320)
Dworkin D. (VII-232)
Dyrenkov S.A. (XII -114)
Dzenis Z.Y. (VI-84)
Dziewonski K. (XI-82; XI-105)
Dzhkeli K.G. (V-90)

E

Eberman F. (VIII-258)
Efrat E. (VIII-268)
Egorov B.N. (II-54)
Elpatyevsky P.V. (V-62)
Elsasser H. (VI-96)
Ena A.V. (V-82)
Ena V.G. (V-82)
Enyedi Gy. (XI-47; VI-201)
Epshtein A.S. (VI-113)
Erdeli G. (VII-224)
Eringis K.(XII-268)
Ernandes S.J. (I-276)
Eropkina N.D. (VI-109)
Ershov A.T. (II-37)
Estes J.E. (II-234)
Evseeva L.S. (II-37)
Evteev O.A. (VII-116)
Eyubov A.D. (V-86)

F

Fairbairn K.J.(VIII-84)
Fakhfakh M.(VII-259)
Farhang Asad A.(IV-118)
Faustova M.A. (I-333)
Federico Sulroca (XII - 259)

Dutt G.K. (VIII-265)
Dutt A.K. (XII -205)

Fedorova T.J. (VII-261)
Fedorovitch B.A.(I-273; I-174;
I-23)
Fenelon P. (I-144)
Fernandez Veiga N. (XII -43)
Ferro G. (IX-17)
Fichtnerova J. (VIII-270)
Filosofov V.P. (I-262)
Findeisen G. (X-24)
Finko E.A. (I-23; I-276)
Flouriot J. (VIII-119)
Fomin G.N. (XI-109)
Ford D.C. (I-46)
Förster H. (X-22)
Fotiev S.M. (I-289)
Fraser D.A. (XII -118)
Frecaut R. (II-181)
Freeberne G.D.M. (VII-118)
Freitag U. (II-323; XII -206)
Fridland V.M. (IV-160)
Friganović M. (VII-29)
Frolov Yu.S. (X-89)
Fuchs R.G. (VII-22)
Fujiwara K. (III-67)

G

Gachechiladze R.G. (VII-264)
Gale S. (VI-27)
Galibert G. (I-146)
Galkina T.A. (VIII-225)
Galkovich B.G. (IX-20)
Gallusser W.A. (VIII-275)
Gambi L. (XII -211)

Fedina A.E. (V-90)
Fedorova N.M. (XII-118)
Gangloff P. (I-332)
Garanin L.K. (V-62)
Garcia A. (VI-130)
Gardavsky V.(XII-209)
Garnier B.J. (II-42)
Gaspar J. (XII-271)
Gassaway A.R. (VII-25)
Gâstesku P. (II-325)
Gavrilova S.A. (X-116)
Gellert J.F. (I-268)
Gelovani V.A. (VIII-234)
Gentilcore R.L. (IX-50)
Geatileschi M.L. (VII-33)
Geodakian R.O.(V-35)
Georgiev T. (V-40)
Gerasimov I.P. (I-49;
            VIII-223; XI-30)
Gerasimova T.P. (X-II)
Gershanovich D.E. (XII-73)
Gil E. (V-119)
Gilewska S. (I-27)
Giurcaneanu C. (VII-118)
Glazov M.V.(XII-125)
Glasovskaya M.A. (V-44)
Glazatscheva L.I. (II-212)
Glazyrin G.E. (II-327)
Gluskova V.G. (VII-266)
Gober-Meyers P. (VII-39)
Gohstand R. (IX-53)
Goica N.I. (V-59)
Gokhman V.M. (VIII-24;
            XI-112; VI-29)
Gold J.R. (VII-269)
Goldenberg L.A. (IX-20)
Golte W. (IV-17)

Gams I. (I-153)
Ganeshin G.S. (I-265)
Goltz G.A. (IX-23)
Golubev G.N. (II-288;
            II-339)
Gonzalez Clemente E. (I-271)
Gonzalez I. (III-15)
Gonzalez M. (VI-92)
Gorkin A.P. (VI-29)
Gorlenko I.A. (XI-71)
Gorodetskaya M.E. (I-273;
            I-23)
Gornung M.B. (VIII-279)
Gorovoy V.L. (VIII-284)
Gorunovich S.A. (VII-213)
Gorunovich V.A. (VII-213)
Goryainova I.N. (IV-23)
Gosal G.S. (VII-120)
Grachev A.F. (I-89)
Gracovish V.F. (II-328)
Gradus Ye. (VIII-120)
Gramoteyeva L.I. (XI-60)
Granberg A.G. (XI-53; XI-143)
Gratsianskij A.N. (X-122)
Grave M.K. (I-273)
Graves N.J. (X-70)
Grebenchikov O.S. (IV-19)
Green A.M. (II-238; XI-134)
Gregory K. (I-158)
Gregori Enrico F. (XII-286)
Grichuk V.P. (I-293;I-333)
Grigoruev S.V. (VI-270)
Grimm F. (VII-272)
Grishanova A.G. (VII-186)
Grossman D. (VIII-29)
Grunberg G.Yu.(X-19)
Gubonina Z.P. (I-376)

Ion Iordan (VIII-135)
Iordanova M. (II-200)
Isachenko A.G. (V-95)
Isakov Y.A. (IV-77)
Isalque S. (I-276)
Isard W. (XI-65)
Ivan A. (XII-24)
Ivanov Yu.N.(II-327)
Ivanova T.D. (VII-170)
Ivantsova G.V. (VI-127)
Ives J.D. (XI-17)

J

Jackson W.A.D. (VIII-136)
Janelle D.G. (IX-56)
Janez Feito H. (X-107)
Jänckel R. (VII-275)
Jaoshvili V.Sh. (VII-173)
Jekouline V.S. (IX-59)
Jenkin J.J. (IV-185, IV-164)
Jerchov E.B. (IX-23)
Jerczynski M. (XI-105)
Jogi J.O. (II-51)
Johannessen C.L. (IV-74)
Joliffe I.P. (III-19)
Jordanov T. (XII-171)
Josan N. (I-173)
Jurchenko V.V.(VIII-234)

K

Kabanas L. (VI-92)
Kadar L. (I-302)
Kagan V.A. (VI-292)
Kaizuka S.(I-52)
Khotinsky N.A. (I-394)
Khrustchev A.T. (VI-105,
                VIII-171)

Kaletskaya M.S. (I-23)
Kalinine A.M. (XII-26)
Kalinina V.R. (V-108)
Kalinova M. (II-372)
Kamasundara B. (II-323)

Kamozawa I. (VI-102)
Kanayev L.A. (II-329)
Kanev D.D. (I-56)
Kaniss Ph. (XI-65)
Kantsebovskaya J.V. (VI-60)
Kansky K.J. (VIII-34)
Kapitsa A.P. (V-62)
Karackashev Ch. (VII-190)
Karan P.P. (VII-277)
Karaska D. (XI-147)
Karinen A.E. (VII-58)
Karing P.H. (II-51)
Karpov L.N. (VI-295, VIII-139)
Kazanskaya N.S. (IV-77)
Kazakov A.J. (VII-186)
Kasumov R.M. (VIII-143)
Katona S. (VII-279)
Kayane I. (II-331)
Kes A.S. (I-273, I-174)
Khailov K.M. (III-10)
Khairoulline R.R. (II-57)
Khalcheva T.A. (I-376)
Kharitonov V.M. (VII-175)
Khmeleva N.V. (I-188)
Khodakov V.G. (I-338)
KhodashovaK.S. (IV-109, XI-130)
Khamjanova N.V. (II-208)
Khodzaev D. (XI-121)
Khorev B. (XI-121)
Korzhuev S.S. (I-49; I-23)
Kosakovskaya N.A. (X-142)
Koshechkin B.I. (I-341)

Kibalchich O.A. (VI-109)
King R.H.(IV-166)
Kinzel H. (X-22)
Kipnis B.A. (VII-282)
Kiradjiev S. (VII-49)
KirillovaT.V. (II-54)
Kistanov V.V. (VI-113)
Klemencic V. (VIII-146)
Kligue R.K. (II-214,
           III-74)
Klimek K. (XII-276)
Kluge K. (XII-278)
Klykov L.V. (XI-162)
Klyukanova I.A. (II-343)

Kobakhidze E.D. (VI-117)
Kobec N.V. (I-255)
Kogutowicz M.G. (VII-126)
Kolaja J. (VIII-151)
Kolobov N.V. (II-57)
Kolosova Yu.A. (VIII-153)
Kolotievsky A.M. (VIII-37,
               VI-84)
Komar I.V. (VI-19, VI-39)
Kondakova L.P. (I-61)
Konovalov G.S. (II-185)
Kopilov V.A. (VII-226)
Korcelli P. (VIII-157)
Korinskaya V.A. (X-11)
Koronkevich N.I. (II-254;
                II-248)
Korsh A.V. (VI-177)
Korsun V.I. (II-334)
Kosinskiy L. (XI-87)
Kosov B.F. (I-188)
Kostrowicki A.S. (IV-82)
Kostrowicki J. (XI-50;
              VI-211)

Kostyaev A.G. (I-343)
Kotlova Z.F. (VII-116)
Kotlyakov V.M. (II-339;
              II-334)
Kortus B. (VIII-159)
Koryakin V.S. (II-298)
Kouprianova T.P. (V-99)
Kovačik Ch.F. (IX-64)
Kovalev S.A. (VII-350)
Kovalevskaya M.K. (X-11)
Kovalskaya N.Ja. (VII-113)
Kovda V.A. (IV-170)
Köves J. (X-28)
Kozakevitch V.P. (IV-149)
Kozuharova A. (VII-285)
Krakover S. (VIII-120)
Krajiček L. (VI-121)
Krajko G. (VI-124)
Kral V. (XII-31)
Kramm H. (VII-54)
Krasilnikova N.V. (X-80)
Krasnopolsky A.V. (VI-177)
Krauklis A.A. (V-27)
Kravtsova V.I. (II-339)
Krenke A.N. (II-301;
Krikounova S.V.(VII-186)
Krivolutsky A.E.(V-47)
Krivolutsky D.A.(IV-85)
Krönert R. (VII-272)
Krotov V.A. (VI-127)
Kruglova G. (VIII-296)
Krylova N.S. (X-142)
Kühnl K. (VII-54)
Kuklinski A. (VIII-163)
Kulikov A.V. (I-406)
Kvasov D.D. (I-3o4)
Kupzov A.J. (IV-87)
Kuskov A.P. (I-62)

Nikolayev V.A. (V-47)
Nikolaeva T.V. (I-89)
Nicolov B. (V-40)
Nikolov F.(VI-144)
Nikonov A.A.(I-358)
Nitz B. (VIII-258)
Nömmik S.J. (VI-53)
North R.N. (VIII-181)
Nosova L.M. (XII -138)
Novak V. (XII -165)
Novakova-Hribova B. (XII -232)

O

Odell P.R. (VI-247)
Odesser S.V. (VIII-225)
Oelke E. (VII-275)
Ofomata G.E.K. (I-196)
Ogoundana B.(VIII-184)
Oguntoyinbo J.S. (II-114)
Ohmura A.(II-80)
Ojany F.F. (I-69)
Ojo O. (II-84)
Olivera M. (VII-132)
Olyunin V.N. (I-23)
Ominde S.H. (VII-67)
Omoto K. (III-67)
Onesti L.J. (I-200)
Orbera H.L. (I-92; I-91)
Orfanov J.K. (VII-189)
Orme A.R. (I-73)
Ortiz-Alvarez M.I. (VII-137)
Ota I. (VII-140)
Otoc S. (VII-143)
Otterman J. (II-168; IV-121)
Outcalt S.I. (XII -52)
Ozenda P. (XII -143)

P

Paczos S. (II-72)
Pagetti F. (VII-159)
Pagney P. (II-90)
Pakalnis R. (XII -268)
Pakhomov M.M. (I-358)
Pal A.K. (I-203)
Palagiano C. (I -397)
Palamarchuk A.M. (VI-320)
Palamarchuk M.M. (VI-320;
                         XI-71)
Palanivelu C. (VI-238)
Palienko V.P. (I-100)
Panaite L. (VII-243)
Pancheshnikova L.M. (X-11)
Panfilov D.V. (IV-96)
Panizza M. (I-279)
Pankova E.I.(IV-192)
Panov M. (VII-305)
Papp-Váry Á. (VI-324)
Paraschiv D. (I-77)
Paratore E. (VI-56)
Parishkura S.I. (I-374)
Paskoff R.P. (I-207)
Paulukat I. (VII-313)
Pavlic Z. (VII-68)
Pecora A. (XII -286)
Pécsi M. (I-78; I-362)
Péguy Ch.P.(II-355)
Pena A. (II-93)
Penkov I. (VII-190)
Pentchev P. (II-359; II-372)
Perelman A.I. (V-44; XII -98)
Perera N.P. (IV-174)
Perera P.D.A. (IV-101)
Perez A. (IV-177)
Perez H.A. (XII -34)

Richling A. (V-140)
Richter N.G. (VIII-317)
Ricour-Singh F. (VI-330)
Rieguelme deRejon D. (II-119)
de la Riva P.J. (IX-27)
Rivera G. (III-19)
Rivero R.F. (I-125)
Robertson J.D. (XII -41)
Roder I.D.
Roder W. (VII-75)
Rodman L.S. (XII -146)
Rodriguez O. (XII -34)
Roesseler B.(VII-310)
Rogers A. (VII-75)
Rogge J. (VII-82)
Romsa G.(VIII-287)
Roseiszewski M.(IX-13)
Rosell D.Z.(VIII-191)
Rosenfeld Ch.L.(I-281)
Rosenvall L.A.(VI-239)
Rosin M.S.(VI-242)
Rosing K.E.(VI-247)
Rosly I.M.(I-87)

Roubina E.A. (I-252)
Roubitschek W. (VII-145)
Roucloux I.C. (XII -239)
Rubinshtein E.S. (II-123)
Rudenko L.G. (VII-116)
Rudneva E.N. (XII -149)
Rump H. (II-258)
Rumpf H. (VII-313)
Runova T.G. (XI-71)
Ruocco D. (VI-334; VI-226)
Rybin G.B. (VII-193)
Rychagov G.I. (I-350)
Ryumin A.K. (I-89)

S

Saavedra G. (IV-177)
Saint-Giron M.Ch. (IV-57)
Saint Moulin de L. (VIII-
192)

Sakaly L.I. (II-65)
Sakharova R.I. (I-252)
Salgueiro T.B. (VII-316)
Salimov H.S. (VIII-193)
Salinas Chavez E. (VIII-246)
Saliyev A. (VII-307)
Salita D.C. (X-131)
Salnikov S.S. (III-56)
Samarina N.N. (II-17)
Samoilenko V.S. (II-37;III-30)
Sandru I. (VII-87)
Sapozhnikov V.V. (II-261)
Saratka G. (VIII-258)
Sasaki H. (VII-319)
Saunders P.F. (I-271)
Saushkin Yu.G. (VIII-65;
VI-39; XI-74)
Savenko V.S. (II-193)
Sazonov N.V. (VIII-234)
Scaramellini G.(VIII-103)
Schafer A. (IV-49)
Scherf K. (VII-310)
Schulz H. (VIII-258)
Schulz M. (VII-313)
Schultz S.S. (I-89)
Schlimme W. (X-51)
Schmidt H. (VI-140)
Schmidt U. (VII-202)
Schmidt R. (IV-179)
Schmithüsen J. (IV-123)
Scholz D. (VI-140)
Scholz E. (I-268)

Schwabe E. (VII-205)
Schytt V. (II-301)
Sdasuk G.V. (VIII-139;
                XI-77)
Segre A. (VII-253)
Seifried N.R.M. (VIII-320)
Selariu O. (III-33)
Selezneva W.S. (V-108)
Semashko I.N. (XI-162)
Semenchenko B.A. (III-30)
Semevski B.N. (X-89; VI-19)
Sergin S.Ya.(II-128)
Serniotti P. (VIII-97)
Seselgis K. (VII-206)
Ševčik F. (XII -242)
Shabad T. (X-133)
Shafi M. (XII -181)
Shamanayev Sh.Sh. (II-263)
Shariguin M.D. (VI-109)
Sharma I.P. (VII-364)
Sharma H.S. (I-219)
Sharma R.C. (III-59;VII-210)
Sharma S.K. (VIII-324)
Sharon D. (II-96)
Shaw D.J.B. (IX-79)
Shcherbakov F.A. (I-61)
Shear J.A. (II-97)
Shein V.S. (I-91; I-92)
Shcherbau M.I. (V-59)
Scherban M.I.(II-65)
Shevnin A.N. (II-208)
Shevtsova N.A. (VII-360)
Shikhlinski E.M. (II-132)
Shilov M.P. (IV-149)
Shinde S.D. (VIII-195)
Shirinov N.Sh. (I-20)
Shlikhter S.B.(VIII-168)

Shnitnikov A.V. (II-308;II-217)
Sholokhov V.V. (I-61)
Shreiber B.T. (XI-162)
Shulenin Yu.P. (XI-57)
Shvets G.E. (II-219)
Siemens A.H. (IX-83)
Silvan J. (VIII-202)
Simon I. (VI-337)
Simonov Yu.G. (I-235; V-62)
Singh B. (VIII-205)
Singh J. (VII-370)
Singh K.N. (I-283)
Singh R.P.B. (VII-88;
Singh R.B. (VII-364)
Singh R.L. (VII-88)
Singh R.P. (I-94; I-336)
Singh R.S. (VIII-53)
Singh S.K. (VIII-195)
Singh U. (VIII-208)
Singh R.Y. (VII-367)
Singh V.R. (VIII-53)
Sinha S.P. (VII-324)
Šipka E. (XII -246)
Sircar P.K. (IX-85)
Sirenko N.A. (I-374; I-368)
Skochan L. (VIII-327)
Skornyakov V.A. (V-47)
Skriabina A.A. (XII-107)
Skublova N.V. (I-255; I-98)
Skulkin V.S. (V-62)
Slater P.B. (XII -250)
Slouka A.E. (VII-90)
Smailes A.E. (XI-82)
Smidovitch I.N.(X-35)
Smidovitch S.G. (VII-186)
Smirnjagin L.V. (VI-29)
Smirnova V.M. (VII-189)

329

Smith D.M. (XI-155)
Smith K.E. (VIII-75)
Smotkine H. (VI-165)
Snytko V.A. (V-27)
Snytkov A. (V-143)
Sochava V.B. (V-27)
Sokolov A.A. (XI-36)
Sokolov I.A. (IV-182)
Sokolov V.A. (X-135)
Sokolovsky I.L. (I-100)
Solntsev V.N. (V-30)
Soloviev S.L. (III-26)
Soloviev V.V. (I-103)
Solovyeva M.G. (X-35)
Sommer M. (X-140)
Sorochinskaya-Goryunova I.I.
              (VII-64)
Speight J.G. (V-126)
Spengler R. (II-267)
Spinelly G. (VIII-331)
Spizhankov L.I. (VII-213)
Šprincova S. (XII -253)
Stadelbauer J. (VI-250)
Stafford H.A. (VIII-334)
Staluppi G. (VII-159; VI-276)
Starkel L. (I-27; V-119; XII
Stasiak J. (I-400)   -276)
Stavrova N.I. (XII -138)
Stefanescu I. (VII-152)
Stepanov V.N. (X-142)
Stigliano M. (XII -256)
Stillwell H.D. (V-146)
Stoddart D.R. (III-93)
Stone K.H. (VII-363)
Strelkov S.A. (I-341)
Strenz W. (IX-88)
Stroev K.F. (XII -306)

Subbotin A.I. (II-369)
Suetova I.A. (III-10)
Sugar T. (XII -276)
Sukroongreung Ch. (III-34)
Suslov V.F. (II-327;
              II-304)
Suzuki T. (I-223)
Svanidze G.G. (II-221)
Svarichevskaja Z.A. (I-89,
                      I-106)
Sverlova L.I. (XII -64)
Swizewski A. (X-49)
Swizewski C. (X-49)
Symader W. (II-258)
Sysueva L.V. (V-108)
Szilard J. (I-355)
Szupryczyński J. (II-270)

T

Taborisskaya I.M. (VII-198)
Tagounova L.N. (XII -132)
Takahashi K. (I-223)
Takamura H. (II-331)
Talskaya N.N. (I-252)
Tanczos-Szabo L. (VII-93)
Tanioka T. (IX-35)
Taramaeva T.A. (VII-213)
Tarasov G.L. (VI-134)
Targulian V.O. (I-402;
              IV-182)
Tarmisto V.Yu. (VIII-213)
Tata R.J.(VIII-55)
Tatai Z. (VI-169)
Tatevosov R.V. (VII-64)
Taylor D.R.F. (IX-92)
Taylor P.K. (X-25)

Vinogradov B.V. (IV-133)
Vinogradov O.N. (II-379)
Vitko K.P. (XII -152)
Vitvitskij G.N. (II-135)
Vivian R. (II-288)
Vivó-Escoto J.A. (I-112;
                  II-101)
Vladimirov V.V. (VI-183;
                  XI-57)
Vlasova T.V. (X-97)
Volevakha N.M. (V-59.)
Volkov N.G. (I-100)
Volobuev V.R. (IV-188)
Voracek V. (VIII-340)
Vornovitskiy M.Ja. (XI-162)
Voronov A.G. (IV-138)
Voskresenski K.P. (II-225)
Voskresensky S.S. (I-235)
Vosovik Yu.(XII-292)
Vostokova E.A.(IV-53)
Vriser I. (VII-335)
Vychivkin D.D. (IV-53)
Vygodskaya N.N. (V-62)

W

Wahla A. (XII -310)
Walker H.J. (III-41)
Warntz W. (VI-348)
Wasilewski L.I. (VI-351)
Weihaupt G.I. (I-238)
West N. (VI-258)
Wheeler J.O. (VII-339)
White G. (XI-11)
Wirth G. (VI-262)
Wishiewski E. (I-240)
Woldenberg M.I.(XII-185)
Wolf M.B. (VI-242)

Wongtangswad N. (I-116)
Woo M.K. (II-204)
Wood J.D. (X-101)

Y

Yamaoka M.(XII-88)
Yanez Feito(X-107)
Yazawa T.(II-138)
Yeates M. (VII-342)
Yegorov V.V. (IV-192)
Yonekura J. (VII-346)
Yoshida Y. (III-67)
Yoshino M.M. (II-103)
Yuill R.S. (VI-326)
Yurchenco V.V. (VIII-234)
Yusin G.S. (VII-294)

Z

Zaika V.E. (II-219)
Zaitsev G.A. (V-62)
Zamkov O.K. (VIII-344)
Zapletal L.A. (XII -37)
Zapletalova J.·(VIII-347)
Zaslavsky M.N. (I-188)
Zaumseil L. (VIII-258)
Zavelsky F.S. (I-379)
Zavoianu I. (II-325)
Zele R.V. (XI-65)
Zeremski B.M. (I-286)
Zezina O.N. (III-46)
Zhachkina S.V. (I-262)
Zhivago A.V. (I-49;III-99)
Zhukov V.T. (VI-307)
Ziapkov L. (V-51)
Zietara T. (I-244)
Zimm A. (VII-313)

CONTENTS

III. GEOGRAPHY OF THE OCEAN

IV. BIOGEOGRAPHY AND SOIL GEOGRAPHY

VIII. REGIONAL GEOGRAPHY

IX. HISTORICAL GEOGRAPHY

Внешторгиздат. Зак.№К125
Тип. №9 Зак. №1928